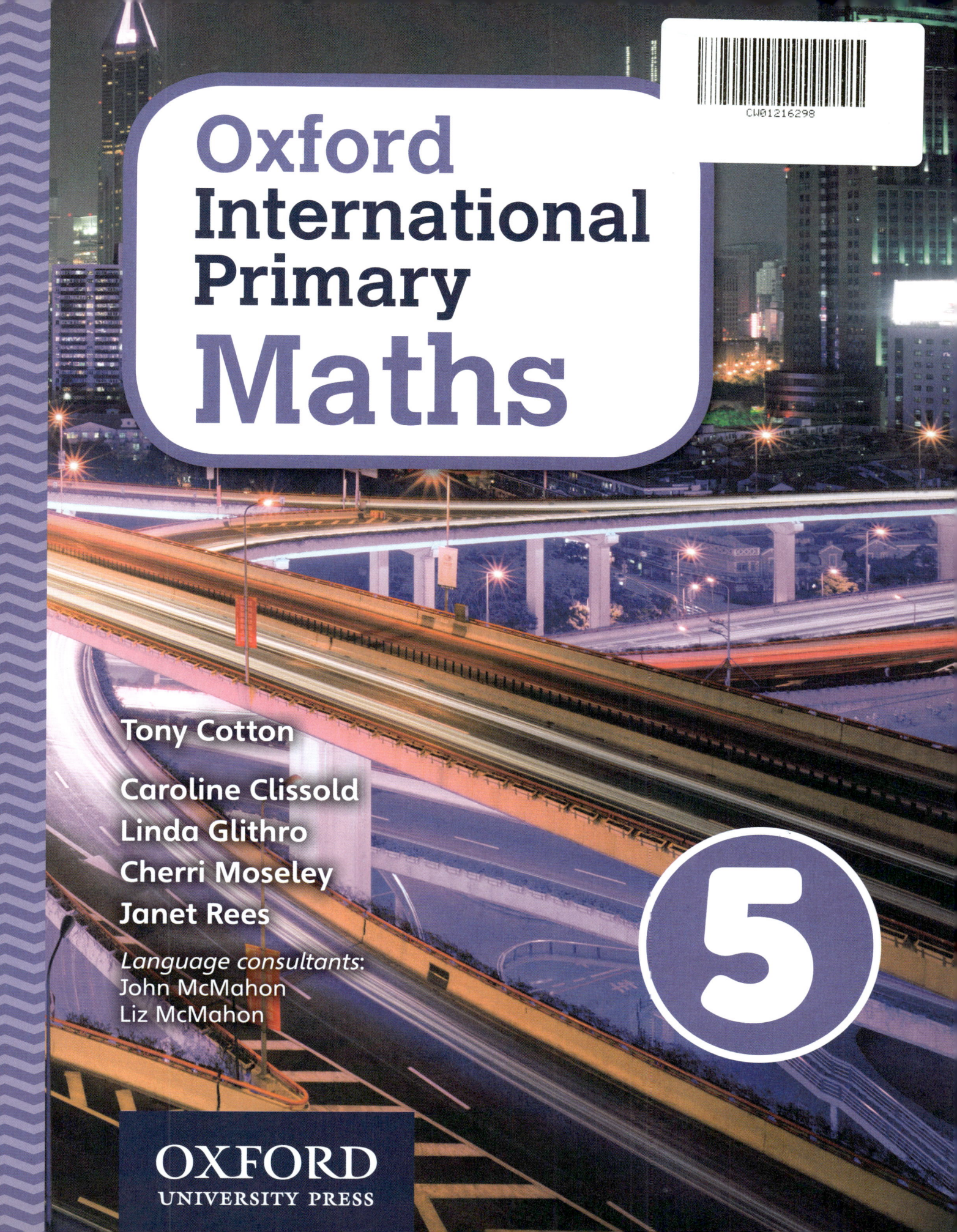

UNIVERSITY PRESS

Great Clarendon Street, Oxford, OX2 6DP, United Kingdom

Oxford University Press is a department of the University of Oxford.
It furthers the University's objective of excellence in research,
scholarship, and education by publishing worldwide. Oxford is a
registered trade mark of Oxford University Press in the UK and in
certain other countries

© Caroline Clissold 2014

The moral rights of the authors have been asserted

First published in 2014

All rights reserved. No part of this publication may be reproduced,
stored in a retrieval system, or transmitted, in any form or by any
means, without the prior permission in writing of Oxford University
Press, or as expressly permitted by law, by licence or under terms
agreed with the appropriate reprographics rights organization.
Enquiries concerning reproduction outside the scope of the above
should be sent to the Rights Department, Oxford University Press,
at the address above.

You must not circulate this work in any other form and you must
impose this same condition on any acquirer

British Library Cataloguing in Publication Data
Data available

978-0-19-839463-1

3 5 7 9 10 8 6 4

Paper used in the production of this book is a natural, recyclable
product made from wood grown in sustainable forests.
The manufacturing process conforms to the environmental
regulations of the country of origin.

Printed in China

Acknowledgements
The publishers would like to thank the following for permissions to
use their photographs:

Cover photo: Martin Puddy/Stone/Getty Images, P111a: M. Unal
Ozmen/Shutterstock, P111b: Peter Cox/Shutterstock, P111c:
Shutterstock, P111d: Roman Sigaev/Shutterstock, P111e: Josep M
Penalver Rufas/Shutterstock, P111f: Johan Swanepoel/Shutterstock,
P111g: Shutterstock, P111h: Patricia Hofmeester/Shutterstock,
P111i: Shutterstock, P111j: Shutterstock, P111k: Shutterstock, P111l:
Shutterstock, P111m: REX/Image Broker, P111n: Shutterstock, P111o:
Shutterstock, P111p: Evgeniy Ayupov/Shutterstock.

Although we have made every effort to trace and contact all
copyright holders before publication this has not been possible in all
cases. If notified, the publisher will rectify any errors or omissions at
the earliest opportunity.

Links to third party websites are provided by Oxford in good faith
and for information only. Oxford disclaims any responsibility for
the materials contained in any third party website referenced in
this work.

Contents

Unit 1	**Number and Place Value**	1
	Engage	
	1A Place value	2
	1B Tenths and hundredths	4
	1C Rounding	6
	1D Ordering and comparing	8
	1E Number sequences	10
	1F Odd and even numbers	12
	Connect	16
	Review	17

Unit 2	**Fractions, Decimals, Percentages, Ratio and Proportion**	19
	Engage	
	2A Equivalent fractions	20
	2B Improper fractions	22
	2C Fractions and decimals – tenths	24
	2D Hundredths	26
	2E Percentages	28
	2F Proportion	30
	2G Ratio	32
	Connect	34
	Review	35

Unit 3	**Mental Calculation Strategies**	37
	Engage	
	3A Number pairs	38
	3B Multiplication and division facts	41
	3C Properties of numbers	44
	3D Counting on and back	46
	3E Near multiples	48
	3F Which strategy?	51
	3G Multiplication strategies	53
	3H More multiplication strategies	55
	3I Doubling and halving	57
	Connect	59
	Review	60

Unit 4	**Written Calculation**	61
	Engage	
	4A Addition and subtraction	62
	4B More addition and subtraction	64
	4C Multiplication	66
	4D Multiplying decimals	69
	4E Division	71
	4F More division	74
	4G Brackets	77
	Connect	80
	Review	82

Unit 5	**Shape**	83
	Engage	
	5A Triangles	84
	5B Symmetry	86
	5C 3D shape	88
	5D Lines	91
	5E Angles	93
	5F Polygons	96
	Connect	98
	Review	100

Unit 6	**Position and Movement**	101
	Engage	
	6A Coordinates	102
	6B Reflection	104
	6C Translations	106
	Connect	109
	Review	110

Unit 7	**Length, Mass and Capacity**	111
	Engage	
	7A Units of measure	112
	7B Measuring length	114
	7C Centimetres and millimetres	116
	7D Measuring mass	118
	7E Measuring capacity	120
	Connect	123
	Review	124

Unit 8	**Time**	125
	Engage	
	8A Converting between units of time	126
	8B Using the 24-hour clock	128
	8C Reading timetables	131
	8D Calculating time intervals	133
	8E Using calendars	135
	Connect	137
	Review	138

Unit 9	**Perimeter and Area**	139
	Engage	
	9A Understanding perimeter	140
	9B Understanding area	142
	9C Calculating areas and perimeters of rectangles	144
	Connect	146
	Review	147

Unit 10	**Handling Data**	149
	Engage	
	10A Frequency tables and pictograms	150
	10B Bar line graphs and line graphs	152
	10C Line graphs	154
	10D Probability	156
	Connect	158
	Review	159

Glossary 161

1 Number and Place Value

Engage

Work with a partner.

Choose a whole number. This is your **sequence** starting number.

- Make up a sequence of 12 numbers that goes forwards and backwards from your starting number in your chosen step size.
- Write your sequence here:

					Starting number

- Now give your sequence to a partner.
- Ask them these questions:

What is the step size?

Can you add 10 more numbers at the beginning and at the end?

Olivia made a sequence.
It had ten numbers.
The step size was 1.5.
It finished on 7.5.
What was her starting number?

Blade made a sequence.
It had five numbers.
It started on 5 and finished on 65.
What was the sequence step size?

1A Place value

Discover

- Choose one number from each column and make a 6 digit number.
- Write your number here:

100 000	10 000	1000	100	10	1
200 000	20 000	2000	200	20	2
300 000	30 000	3000	300	30	3
400 000	40 000	4000	400	40	4
500 000	50 000	5000	500	50	5
600 000	60 000	6000	600	60	6
700 000	70 000	7000	700	70	7
800 000	80 000	8000	800	80	8
900 000	90 000	9000	900	90	9

- Do this five more times:

- Now order your numbers from smallest to largest:

100 000	10 000	1000	100	10	1

Haris picked 600 000, 4000, 60 and 2.

What number did he make?

- Write it here:

Sofia picked 100 000, 70 000 and 900.

What number did she make?

- Write it here:

1A Place value

Explore

- Make some **5-digit numbers** using your digit cards.
- Write them here:

- Partition your numbers into 10s of thousands, thousands, hundreds, tens and ones:

10s of thousands	
Thousands	
Hundreds	
Tens	
Ones	

- Do this again for **6-digit numbers**:

100s of thousands	
10s of thousands	
Thousands	
Hundreds	
Tens	
Ones	

thousand, forty, sixty, twenty, seven, eighty, nine, thirty, hundred, seventeen, one, eight

- Write these numbers in words:

1. 40 467 _____

2. 89 508 _____

3. 25 049 _____

4. 17 001 _____

5. 234 020 _____

Yukesh made a number.

There were 5 **tens of thousands**, 6 **thousands**, no **hundreds**, 3 **tens** and 2 **ones**.

- Write his number here:

He then swapped the 2 and 5.

Approximately how much smaller is his number now?

- Write his new number here:

1B Tenths and hundredths

Discover

100 000	10 000	1000	100	10	1	•	10th	100th

- Pick three digit cards.
- Place the digits in the grid to make a number.

How many different numbers can you make?

- Write them here:

Nafisat made a 3-digit number with one **decimal place**.

She used the digits 6, 8 and 9.

- Write a possible number she made:

Are there any other possibilities?

- Write them here:

4

1B Tenths and hundredths

Explore

100 000	10 000	1000	100	10	1	.	10th	100th

- Pick four digit cards.
- Make a 4-digit number.
- Place your number in the grid.
- Multiply your number by 10.
- Write the new number here:

- Make another 4-digit number using the same cards.
- Place the number in the grid.
- Divide the number by 100.
- Write the new number here:

- Make another 4-digit number with the same cards.
- Place your new number in the grid.
- Divide this number by 10.
- Write the new number here:

Rashid had a number.

He multiplied his number by 10 and got 134.

What was his original number?

Naomi had a tray of 12 eggs.

How many eggs are there in 100 identical trays?

- Make another 4-digit number with the same cards.
- Place this number in the grid.
- Multiply the number by 100.
- Write the new number here:

Courtney collected 10c coins.

She had $11 of 10c coins.

How many 10c coins did she have?

1C Rounding

Discover

Here is a number line:

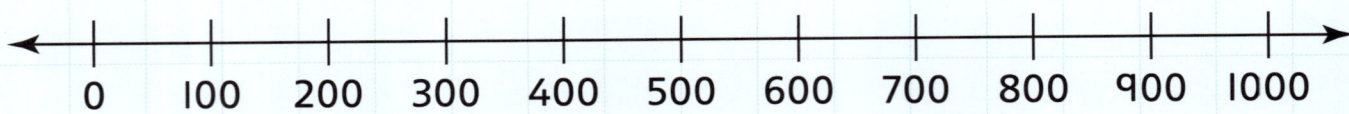

- Estimate where these numbers go on the number line:

  450, 125, 20, 805, 345, 675, 950, 515

- Mark the numbers on the number line.
- Now complete this chart:

	Round to nearest 100	Round to nearest 10	Round to nearest 1
450			
125			
20			
805			
345			
675			
950			
515			

1. Benji saw two books that he wanted to buy.

   The books each cost $19.99.

   He has $40. Is that enough? How do you know?

2. Trudi thought of a number.

   Her number was between 850 and 900.

   She **rounded** it to the nearest 10 and got 870.

   - Write a possible number she thought of:

3. Rasheed thought of a number between 7000 and 8000.

   It was a multiple of 10.

   He rounded it to the nearest thousand and got 7000.

   - Write a possible number he thought of:

1C Rounding

Explore

Play this game to practise your skills of rounding.
- Use a set of 0–9 digit cards.
- Shuffle the digit cards.
- Take it in turn to pick one digit card.
- Find a number in the stars that you can round to the number you picked.
- Cover the star with a counter.
- Continue playing until all the stars are covered.
- The winner is the player with the most stars covered.

1D Ordering and comparing

Discover

| 0 | 1 | 2 | 3 | 4 |
| 5 | 6 | 7 | 8 | 9 |

- Work with a partner.
- Write down ten numbers each. (Write a variety of numbers with a different number of digits. Include both **positive** and **negative** numbers.)
- Cut your numbers out and pile them face down in front of you.
- Take it in turns to pick a number and place it face up on the table.
- As you pick numbers place them in order, from lowest to highest.
- Write the order here:

- Now choose pairs of numbers. Write them with the symbols > or < to show comparison.

  For example: 23.5 < 50   47 > −12

- Write your pairs here:

April had $23.95. Faith had $23.50.

Who had the most money?

- Write the amounts in two number sentences using the < and > symbols:

Greg had 124 shells. Hope had 142 shells.

Who had the most shells?

- Write the amounts in two number sentences using the < and > symbols:

1D Ordering and comparing

Explore

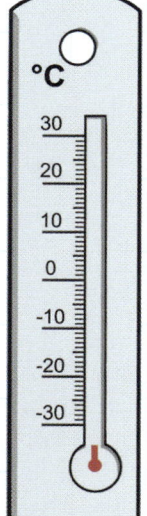 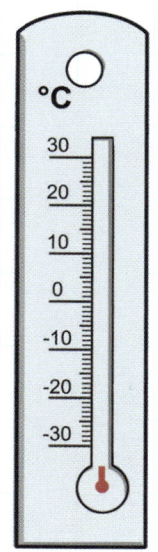 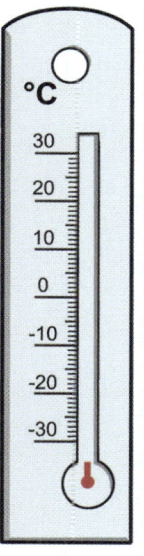 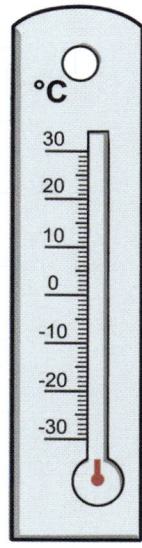

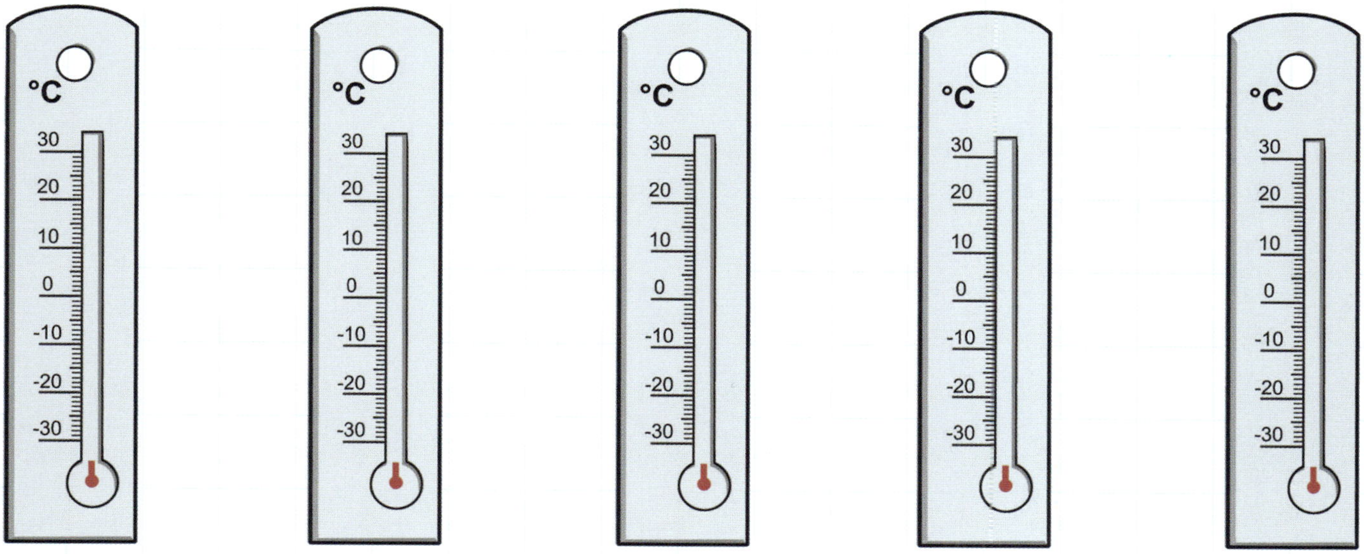

- Place these temperatures on the thermometers:
    1. 30 °C
    2. 10 °C
    3. −25 °C
    4. −5 °C
    5. 24 °C

- Now make up some number sentences using < and >.

    For example: 30 °C > 24 °C.

- Write some statements to match your number sentences.

    For example: 30 °C is 6 degrees hotter than 24 °C.

1E Number sequences

Discover

- Place your set of digit cards face down on the table in front of you.
- Pick three cards.
- Place them in **ascending** order.
- Use them to start two different number sequences with up to ten numbers.
- Don't forget to write the rule!

> For example:
> Digit cards: 3, 6, 7
> Sequence 1:   3, 6, 7, 10, 11, 14, 15, 18, 19, 22   Rule: +3, +1
> Sequence 2:   3, 6, 7, 14, 15, 30, 31, 62, 63, 126   Rule: double, +1

- Write your sequences and rules here:

Digit cards:
Sequence 1:   Rule:
Sequence 2:   Rule:

Digit cards:
Sequence 1:   Rule:
Sequence 2:   Rule:

Digit cards:
Sequence 1:   Rule:
Sequence 2:   Rule:

Digit cards:
Sequence 1:   Rule:
Sequence 2:   Rule:

1E Number sequences

Explore

In a Fibonacci-type sequence each number is the sum of the two previous numbers.

For example: 1, 1, 2, 3, 5, 8, 13…

(3 = 1 + 2   5 = 2 + 3 and so on)

- Extend the Fibonacci sequence for another ten numbers:

- Continue these Fibonacci-type sequences until you pass 100:

    1. 4, 7, 11, 18

    2. 12, 23, 35

    3. 1, 10, 11

- Make up two of your own Fibonacci-type sequences.

    Write them here:

Now for a challenge!

I made up a Fibonacci-type sequence.

It contains 89. It begins with 2.

What is it?

- Write the first 12 numbers here:

1F Odd and even numbers

Discover

Use your interlocking cubes to find out:

1. What happens when you add two odd numbers?

2. What happens when you add three odd numbers?

- Write six examples:

- Write six examples:

Why does this happen?

Why does this happen?

3. What happens when you add four odd numbers?

- Write six examples:

Why does this happen?

- Now, repeat for subtraction with two odd numbers:

three odd numbers:

four odd numbers:

1F Odd and even numbers

Explore

Use your interlocking cubes to find out:

1. What happens when you multiply single digit numbers by 4?

   Do you think that the answers are **odd** or **even** or a mixture of both?

   I think _____

   - Test out your thinking with the cubes.

   Were you right?

   Write some examples:

2. What happens when you multiply single digit numbers by 5?

   Do you think that the answers are odd or even or a mixture of both?

   I think _____

   - Test out your thinking with the cubes.

   Were you right?

   Write some examples:

3. What can you say about:

    a) Multiplying by even numbers?

    b) Multiplying by odd numbers?

1 Number and place value

Connect

- Work with a partner and two sets of digit cards.
- Shuffle the cards and place them face down on the table in front of you.
- Take it in turns to pick a number.
- Choose where to place your digit in the table.
- Write the number it represents in the correct place in the table.
- Continue until you have made a 6-digit number.
- Write your 6-digit number in the last column.
- Rearrange them to make the largest number possible.
- Each time your number is larger than your partner's number you score a point.
- The player with the highest score is the winner.

	100 000	10 000	1000	100	10	1	Number
1							
2							
3							
4							
5							
6							
7							
8							

1 Number and place value

Review

What's my number?

- Think of a 4-digit number.
- Write your number where no one can see it.
- Choose one friend to go first. Ask them questions about their number. They can only answer 'yes' or 'no'.
- Try to ask questions to find the answer as quickly as possible.

Good questions to ask are:

> Is it a multiple of 10?

> Does it round up?

> Is it greater than 4500?

> Is it an odd number?

Your group can ask up to 20 questions.

- Write the numbers that you all thought of here:

- Now order the numbers from smallest to largest:

- Make up some number sentences for these numbers using the < and > symbols.

2 Fractions, Decimals, Percentages, Ratio and Proportion

Engage

$\frac{7}{10}$ $\frac{2}{3}$ $\frac{1}{10}$ $\frac{1}{3}$ $\frac{3}{4}$ $\frac{3}{5}$ $\frac{1}{8}$

- Work with a partner.
- Take two pieces of different-coloured A5 paper.
- Look at them. Are they the same size?
- Now take one piece. Fold the paper and then tear it in **half**.

What **fraction** of the whole piece have you made? _____

What did you divide the piece of paper by? _____

$\frac{3}{10}$

$\frac{3}{8}$

- Now stick the smaller piece in the middle of the larger piece like this:

What fraction is the piece in the middle? _____

What fraction is the piece around the outside? _____

- Cut out the piece around the outside.

What two fractions do you have? _____

- Now cut the outside piece into four strips of the same length.
- Stick these strips onto the other shape.

What do you notice?

$\frac{1}{2}$

$\frac{7}{8}$

$\frac{4}{5}$

Aadil doesn't think that this rectangle is divided into quarters.

What do you think?

Take a piece of A5 paper.

- Divide the paper into four pieces as in the picture.
- Cut out the four pieces.

Are the four pieces equal sizes? _____

Can you prove that these are **quarters**? _____

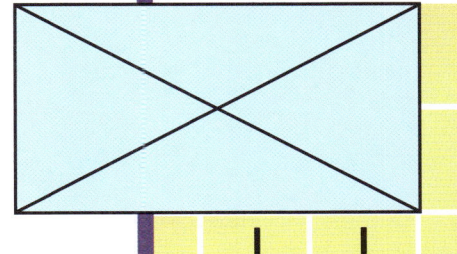

$\frac{1}{4}$ $\frac{1}{5}$ $\frac{2}{5}$

2A Equivalent fractions

Discover

- Take four strips of paper.
- Keep one strip whole.
- Fold the second strip in half.
- Open it up and label each section with the fraction it shows:

| $\frac{1}{2}$ | $\frac{1}{2}$ |

- Fold the third strip into half and half again.
- Open it up and label each section with the fraction it shows:

| $\frac{1}{4}$ | $\frac{1}{4}$ | $\frac{1}{4}$ | $\frac{1}{4}$ |

- Fold the fourth strip into half, half again and half once more.
- Open it up and label each section with the fraction it shows.
- Put your fraction strips in order, largest to smallest, under the first whole strip.
- Draw lines and write labels on this diagram to show the fractions you made:

- Now write down all the **equivalent fractions** that you can see:

Mussaret wants to know how many sixths are the same as $\frac{2}{3}$.

- Use other strips of paper to show her.
- Draw the equivalent fractions on this diagram:

- Write some new equivalent fractions that you can see from both diagrams:

2A Equivalent fractions

Explore

- Use your fraction strips and counters from the Discover activity to help you find:

1. $\frac{2}{3}$ of 24

    - Write another fraction that is equivalent to this,

    for example: $\frac{4}{6}$ _____

2. $\frac{5}{6}$ of 24

    Can you think of another fraction and amount that give the same answer?

    For example: $\frac{1}{2}$ of 40 _____

3. $\frac{3}{4}$ of 32

    - Write another fraction that is equivalent to this: _____

4. $\frac{4}{5}$ of 20

    - Write another fraction that is equivalent to this: _____

5. $\frac{3}{10}$ of 40

    Can you think of another fraction and amount that give the

    same answer? _____

6. $\frac{5}{8}$ of 16

    Can you think of three other fractions and amounts that give the same answer?

    _____   _____   _____

7. $\frac{2}{5}$ of 30

    - Write another fraction that is equivalent to this: _____

    Can you think of three other fractions and amounts that give the same answer?

    _____   _____   _____

8. $\frac{4}{8}$ of 40

    - Write two other fractions that are equivalent to this:

    _____   _____

    Can you think of three other fractions and amounts that give the same answer?

    _____   _____   _____

2B Improper fractions

Discover

- Draw a line to match the **improper** fraction with the equivalent **mixed number**.

Mixed numbers (left column): $1\frac{1}{2}$, $1\frac{1}{4}$, $2\frac{1}{4}$, $3\frac{3}{4}$, $1\frac{3}{8}$, $3\frac{5}{8}$, $1\frac{3}{5}$, $1\frac{7}{10}$, $2\frac{3}{5}$, $4\frac{3}{10}$

Improper fractions (middle): $\frac{3}{2}$, $\frac{5}{4}$, $\frac{5}{2}$, $\frac{7}{2}$, $\frac{7}{4}$, $\frac{11}{4}$, $\frac{9}{8}$, $\frac{9}{4}$, $\frac{6}{5}$, $\frac{21}{8}$, $\frac{13}{4}$, $\frac{13}{5}$, $\frac{15}{4}$, $\frac{8}{5}$, $\frac{23}{8}$, $\frac{29}{8}$, $\frac{17}{10}$, $\frac{11}{8}$, $\frac{14}{5}$, $\frac{43}{10}$

Mixed numbers (right column): $3\frac{1}{2}$, $2\frac{1}{2}$, $3\frac{1}{4}$, $2\frac{3}{4}$, $2\frac{7}{8}$, $2\frac{5}{8}$, $2\frac{4}{5}$, $1\frac{1}{8}$, $1\frac{1}{5}$, $1\frac{3}{4}$

A line is drawn from $\frac{7}{2}$ to $3\frac{1}{2}$.

2B Improper fractions

Explore

Follow these steps for each of the statements:
- Change the improper fractions into mixed numbers.
- Find the equivalent fractions to the fraction part.
- Write a number sentence to show what you think.

Is each statement true or false?

For example:

> $\frac{8}{5}$ is less than $\frac{18}{10}$
> This is true because $\frac{8}{5} = 1\frac{3}{5}$, $\frac{18}{10}$ is $1\frac{8}{10}$ or $1\frac{4}{5}$
> $1\frac{3}{5} < 1\frac{4}{5}$

$\frac{7}{4}$ is greater than $\frac{11}{8}$

This is true/false because

$\frac{14}{3}$ is less than $\frac{18}{6}$

This is true/false because

$\frac{23}{5}$ is greater than $\frac{46}{10}$

This is true/false because

$\frac{7}{2}$ is greater than $\frac{26}{8}$

This is true/false because

$\frac{7}{5}$ is greater than $\frac{18}{15}$

This is true/false because

2C Fractions and decimals – tenths

Discover

- Work with a partner
- Use two sets of digit cards.
- Shuffle the digit cards together and place them face down on the table.
- Take it in turns to pick three cards.
- Make a 3-digit number with one decimal place.
- Put your number on the ladder, with the smallest number at the bottom and the largest at the top.
- Put your cards at the bottom of the pile.
- Continue to make new 3-digit numbers with one decimal place.

  When you make a number that can't fit onto the ladder, you score 2 points.
- When you have a full ladder, count your points.

  The player with the lowest score wins. Play the game three times. Who is the winner?

- Write the decimal fractions you made as mixed fractions:

2C Fractions and decimals – tenths

Explore

- Write these numbers on the number line:

$\dfrac{1}{2}, \dfrac{1}{4}, \dfrac{3}{4}, \dfrac{1}{5}, \dfrac{2}{5}, \dfrac{3}{5}, \dfrac{4}{5}, \dfrac{1}{10}, \dfrac{3}{10}, \dfrac{7}{10}, \dfrac{9}{10}$

- Now write the equivalent decimal fractions under the number line.
- Use your number line to answer these problems.

1. Taima thinks that 0.3 is bigger than $\dfrac{1}{4}$.

   Is she correct? How do you know?

   I think she is correct/not correct because _____

   _____

2. Mehtab thinks that $\dfrac{4}{5}$ and 0.6 are equivalent.

   Is he correct? How do you know?

   I think Mehtab is correct/not correct because _____

   _____

3. Chan tried to order some numbers, smallest to largest. He wrote his numbers in this order:

   0.1, $\dfrac{1}{2}$, 0.3, $\dfrac{4}{5}$, 0.6.

   - Write the numbers in the correct order:

   _____, _____, _____,

   _____, _____

   Is Chan correct?

   I think that Chan is correct/not correct because _____

   _____

   _____

4. Naomi thinks that 0.7 of a dollar is more than $\dfrac{3}{4}$ of a dollar.

   Is she correct? How do you know?

   I think she is correct/not correct because _____

   _____

   _____

5. Poppy can choose 0.8 of $10 or $\dfrac{3}{4}$ of $10.

   She wants to choose the largest amount.

   She chooses 0.8 of $10.

   Is this the correct decision? Why?

   I agree/don't agree with her decision because _____

   _____

   _____

25

Fractions, Decimals, Percentages, Ratio and Proportion

2D Hundredths

Discover

- Pick four digit cards and write them in the table.
- Make a **number with two decimal places** and write them in the table.
- **Round** your number to the **nearest whole number**.
- Then round your number to the **nearest tenth**.
- Write your answers in the table.
- Use the same cards to make a second number and write them in the table.
- **Round** your number to the **nearest whole number**.
- Then round your number to the **nearest tenth**.
- Write your answers in the table.
- **Compare** your two numbers using $>$ or $<$.

An example is shown in the table.

How many numbers can you make and round in the time allowed?

Digits	1st number	Rounded to nearest whole number	Rounded to nearest 10th	2nd number	Rounded to nearest whole number	Rounded to nearest 10th	Comparison
7, 6, 3, 1	36.17	36	36.2	71.36	71	71.4	36.17 < 71.36

2D Hundredths

Explore

- Use four digit cards to make numbers with two decimal places.
- Choose for your numbers to be quantities of **metres**, **kilograms or litres**.
- Use a mixture of them all as you work through the activity.
- Put your numbers and amounts in the table.
- Complete the rest of the table by rounding, following the instructions. An example is shown for you.

Numbers picked	Amount made	Round to the nearest 10th	Round to the nearest whole number
6, 4, 9, 2	49.62 kg	49.6 kg	50 kg

2E Percentages

Discover

- Imagine that 100% is $260.
- Write down other **percentages** of this amount.
  How many can you find in two minutes?
- Write your percentages in the diagram. Some examples are shown:

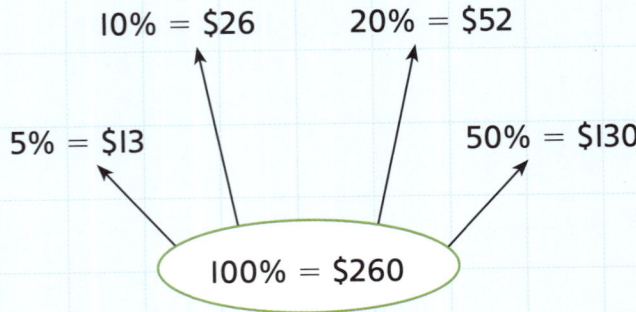

- Use your percentages to answer these problems.

1. My friend bought a TV in a sale.
   Before the sale the price was $260.
   In the sale it was 15% cheaper.
   How much cheaper, in $, is the TV now?

2. My friend bought a coat in the sale.
   Before the sale it cost $260.
   In the sale it was 55% cheaper.
   How much did it cost in the sale?

3. Brian had $260.
   He spent 30% of his money on a laptop.
   How much did he spend?
   How much money did he have after he bought the laptop?

2E Percentages

Explore

- Pick two digit cards and a zero.
- Make a 3-digit multiple of 10.
- Calculate as many percentages of that number as you can in two minutes.
- Write your percentages here:

- Now solve these problems:
  Show how you work them out.

1. There are 100 loaves of bread in a shop. 40% are sliced.
   How many loaves are not sliced?
   How do you know?

2. There were 680 people at a carnival.
   Half of the people were children.
   25% of the children were girls.
   The other children were boys.
   How many girls were at the carnival?
   How many boys were at the carnival?

3. Sam wants to buy a pair of jeans.
   Before the sale the jeans cost $50.
   In the sale the jeans are reduced by 20%.
   How much do they cost now?

4. There are 30 students in Trudy's class.
   40% of the class are boys.
   How many boys are there in the class?
   How many girls are there in the class?

5. Jeans cost $45 before a sale.
   In the sale there is a 20% discount.
   What is the new price of the jeans?

2F Proportion

Discover

1. There are 10 counters.

   6 counters are yellow.
   4 counters are green.

   What **proportion** is yellow?
   Write this using another fraction:

   What proportion is green?
   Write this using another fraction:

2. A pizza is divided into 12 pieces.

   Sami eats three pieces.

   What proportion of the pizza has not been eaten?

   Draw a picture to show this:

3. A bowl contains 12 different kinds of fruit.

   There are 6 kiwi fruit, 4 mangos and 2 bananas.

   What proportion of the fruit are:

   - bananas?   Write this using another fraction:

   - kiwi fruit?   Write this using another fraction:

   - mangos?   Write this using another fraction:

4. A pie is made with bananas and oranges.

   There is a total of 6 kg of fruit in the pie.

   The proportion of bananas is $\frac{2}{3}$.

   How many kilograms of oranges are there?

5. There are 28 students in a class.

   16 of the students are girls.

   What proportion are boys?

   Write your answer in two ways.

2F Proportion

Explore

- Solve these problems.
- You can use drawings or counters to help you.

1. Aran has a collection of toy animals.
   $\frac{1}{2}$ are lions, $\frac{3}{10}$ are monkeys and $\frac{1}{5}$ are giraffes.
   Calculate possible numbers of each animal.
   For example: There are 20 animals in total.

   There are 10 lions, 6 monkeys and 4 giraffes.

   Work out another possibility.

2. India has a bowl of fruit.
   $\frac{1}{4}$ are apples, $\frac{1}{3}$ are papayas, $\frac{1}{6}$ are mangos and the other fruit are bananas.
   Calculate possible numbers of each fruit.
   For example: There are 24 fruit in total.

   There are 6 apples, 8 papayas, 4 mangos and 6 bananas.

   Work out another possibility.

3. Cian has a collection of coins.
   $\frac{1}{3}$ are cents, $\frac{1}{6}$ are 10c coins and the other coins are 25c coins.
   Calculate possible numbers of each coin.
   For example: Cian has 30 coins in total.

   There are 10 cent coins, 5 10c coins and 15 25c coins.

   Work out another possibility.

2G Ratio

Discover

These are the ingredients needed for a fruit salad. It serves 4 people.

2 apples

3 peaches

4 bananas

5 kiwi fruit

1 litre orange juice

The **ratio** of apples to peaches is 2:3.

What is the ratio of:

1. Apples to bananas?

2. Peaches to kiwi fruit?

3. Apples to kiwi fruit?

4. Peaches to bananas?

5. All of the fruit used?

- Rewrite the ingredients needed to serve 8 people:

- Now rewrite the ingredients needed to serve 2 people:

- Use what you know to write the ingredients needed to serve 10 people:

2G Ratio

Explore

1. Hope has 8 bars of chocolate. $\frac{3}{8}$ are dark chocolate. The other bars are white chocolate.

   What is the ratio of dark to white chocolate?
   - Draw a picture to show this.
   - Write the ratio beside your picture.

2. Grace has 20 hair braids. $\frac{3}{4}$ of the hair braid beads are red. The other hair braid beads are blue.

   What is the ratio of red to blue hair braid beads?
   - Draw a picture to show this.
   - Write the ratio beside your picture.

3. Shafi has 15 animals on his farm. $\frac{2}{5}$ of the animals are chickens. The other animals are sheep.

   What is the ratio of chickens to sheep?
   - Draw a picture to show this.
   - Write the ratio beside your picture.

Here's a challenge!

4. Dom has 24 counters. $\frac{1}{3}$ of the counters are blue. $\frac{1}{2}$ are red. The other counters are yellow.

   What is the ratio of blue, red and yellow counters?
   - Draw a picture to show this.
   - Write the ratio beside your picture.

2 Fractions, decimals, percentages, ratio and proportion

Connect

How quickly can you complete this table of equivalences? Time yourself!

Fraction	Decimal fraction	Percentage
$\frac{6}{10}$		
	0.5	
		80%
$\frac{9}{10}$		
		30%
	0.7	
		40%
$\frac{1}{10}$		
	0.2	

- Share your answers with your group. Do you all agree?
- Work as a group to complete this table of equivalences:

Fraction	Decimal fraction	Percentage
$\frac{4}{5}$		
$\frac{1}{5}$		
$\frac{2}{5}$		
$\frac{3}{5}$		
	0.25	
	0.75	
		15%
		82%

2 Fractions, decimals, percentages, ratio and proportion

Review

- Work with a partner.

  What do you now know about fractions, decimal fractions, percentages, proportion and ratio?

- Write your information around the mind map.
- Give examples to support your information.

Some examples are shown.

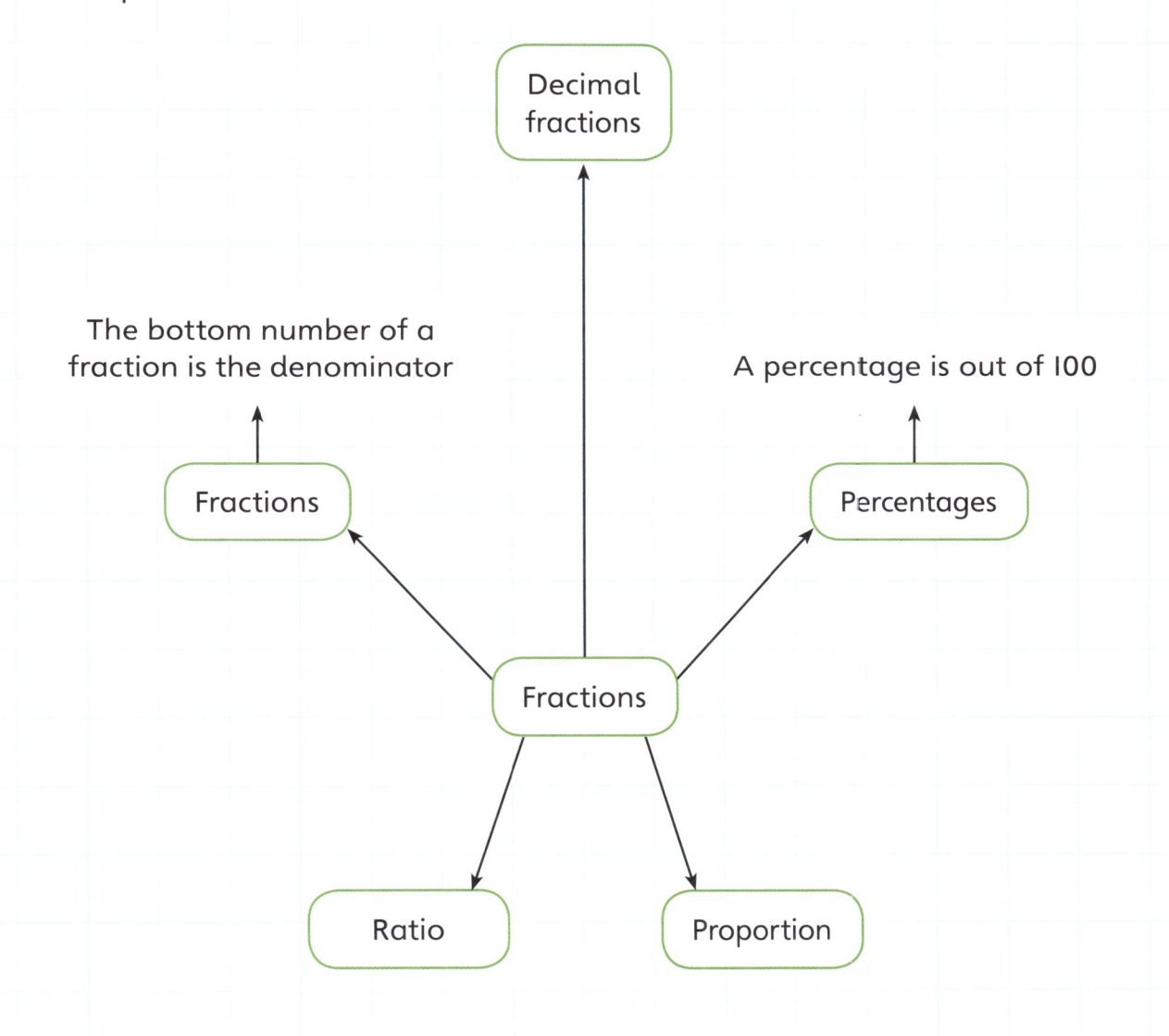

```
                    ┌──────────┐
                    │ Decimal  │
                    │ fractions│
                    └────▲─────┘
                         │
 The bottom number of a          A percentage is out of 100
 fraction is the denominator
         ▲                                    ▲
         │                                    │
    ┌─────────┐                         ┌───────────┐
    │Fractions│                         │Percentages│
    └─────────┘                         └───────────┘
          ▲                               ▲
           \                             /
            \                           /
             \      ┌─────────┐        /
              ─────│Fractions│────────
                    └─────────┘
                    /         \
                   /           \
              ┌───────┐     ┌──────────┐
              │ Ratio │     │Proportion│
              └───────┘     └──────────┘
```

3 Mental Calculation Strategies

Engage

Bobby answered some calculations.

His teacher wasn't pleased!

She said that it was possible to do them faster and more accurately using **mental calculation strategies**.

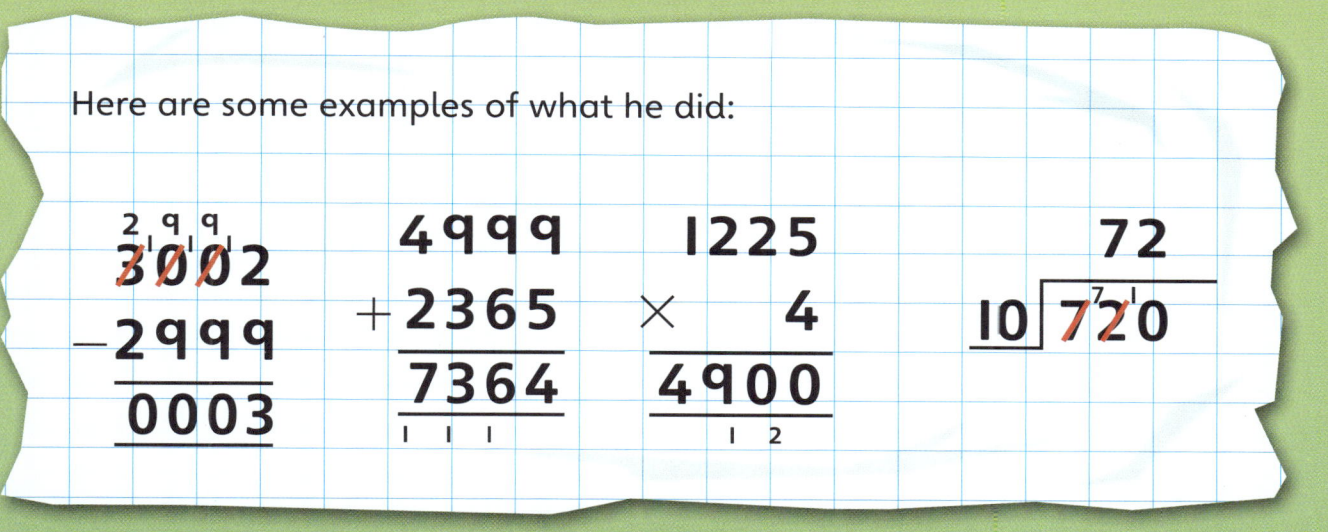

Here are some examples of what he did:

$$\begin{array}{r} \overset{2\,9\,9}{\cancel{3}\cancel{0}\cancel{0}2} \\ -2999 \\ \hline 0003 \end{array} \qquad \begin{array}{r} 4999 \\ +2365 \\ \hline 7364 \\ {\scriptstyle 1\ 1\ 1} \end{array} \qquad \begin{array}{r} 1225 \\ \times\quad 4 \\ \hline 4900 \\ {\scriptstyle 1\ 2} \end{array} \qquad \begin{array}{r} 72 \\ 10\overline{)7\cancel{2}\cancel{0}} \end{array}$$

Can you do these calculations using mental calculation strategies?

- Show Bobby how!
- Use numbers and diagrams that are quicker:

| 3002 − 2999 | 4999 + 2365 | 1225 × 4 | 720 ÷ 10 |

- Now make up some more calculations that you can answer using your strategies.

3A Number pairs

Discover

- Match these number pairs to make 1.
- Draw a line from one number to the number it matches with.
  For example: 0.3 matches with 0.7 to make 1.

0.3   0.8   0.5   0.1   0.2   0.9   0.4   0.6   0.7

0.3   0.8   0.5   0.1   0.2   0.9   0.4   0.6   0.7

- Use your answers to solve these problems:

1. Ci has a length of string that is 12.3 cm. He needs a length of string that is 13 cm.

   How much longer does his string need to be?

2. Mara has 2.4 kg of apples. She needs 3 kg of apples.

   How much more does she need?

3. Sonny cycled 24.8 km. He then stopped for a rest. The ride was 25 km long in total.

   How much longer did he have to cycle?

4. Ben swam 14.1 km. He wanted to swim 15 km.

   How much further did he need to swim?

- Check that your answers make sense.

3A Number pairs

Explore

- Match the number pairs to make 10.
- Draw lines from the numbers on the left to their number pairs on the right. For example: 4.3 matches with 5.7 to make 10.

Left	Right
4.3	3.3
6.7	8.2
2.9	0.4
5.3	6.1
7.7	4.7
3.9	5.7
1.8	7.3
9.6	2.3
2.7	7.1
8.6	1.4

Mental Calculation Strategies

- Use your answers to solve these problems:

1. Nabila was on a 10 km run.

   She stopped for a rest after 2.7 km.

   How much further did she have to run?

2. Tareq has $8.60.

   He needs $10 to buy a pair of shorts.

   How much more does he need?

3. A snail travels 10 cm in total.

   It travels 3.9 cm in one minute.

   How much further does the snail need to go?

4. Samir is filling a 10 l bucket with water. He fills it to 5.3 l.

   How much more water does he need to fill the bucket?

- Check that your answers make sense.

3B Multiplication and division facts

Discover

How quickly can you answer these calculations?
Ask a friend to time you.

1. 8 × 4 =	5. 42 ÷ 6 =	9. 63 ÷ 7 =	13. 16 ÷ 4 =
2. 7 × 3 =	6. 4 × 9 =	10. 3 × 6 =	14. 25 ÷ 5 =
3. 54 ÷ 9 =	7. 30 ÷ 3 =	11. 8 × 7 =	15. 8 × 8 =
4. 3 × 9 =	8. 6 × 8 =	12. 30 ÷ 6 =	16. 81 ÷ 9 =

Time taken:

- Use your knowledge of **multiplication and division facts** to answer these problems:

  1. Khidra scored 30 runs in a cricket match.

     Qasif score six times as many runs.

     How many runs did Qasif score?

  2. Nyla has a collection of 80 stamps.

     Her friend has five times as many stamps.

     How many stamps does her friend have?

Mental Calculation Strategies

3. Jiao can write her name 20 times in a minute.

   Sue can write her name three times as many as Jiao in a minute.

   How many times can Sue write her name in a minute?

4. Paul has a collection of 150 toy cars.

   He divides them equally into five piles

   How many cars are in each pile?

5. Sammi has 240 counters.

   She divides them equally into eight piles.

   How many counters are in each pile?

6. Paulo has a collection of 20 seashells.

   Freddie has nine times as many seashells.

   How many shells does Freddie have?

3B Multiplication and division facts

Explore

- Work with a partner.

You need a set of digit cards with an extra zero card.

- Shuffle the digit cards and place them in a pile on the table.
- Decide who will start.
- The first player chooses three cards and makes a 3-digit number.
- Try to make a number that can be divided by one or more of the numbers 2, 5, 10 and 100.

You score:

- 1 point if you can divide it by 2
- 2 points if you can divide it by 5
- 3 points if you can divide it by 10
- 4 points if you can divide it by 100
- Take it in turns to make numbers.
- Record your results in this table:

Player 1 Name:		Player 2 Name:	
Number	Score	Number	Score
Total score		Total score	

Who won?

3C Properties of numbers

Discover

- Work with a partner.
- Place the numbers below in the correct position in the table. Only one number can go into each section!

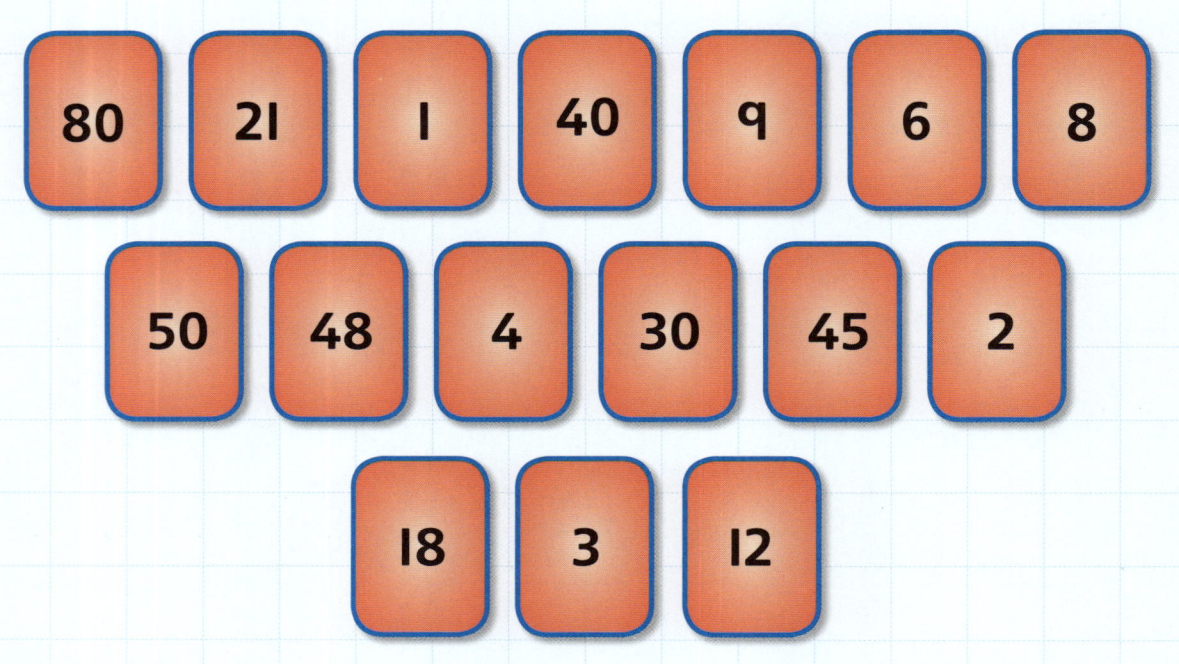

80, 21, 1, 40, 9, 6, 8, 50, 48, 4, 30, 45, 2, 18, 3, 12

	Multiple of 3	Factor of 24	Multiple of 5	Factor of 36
Even number				
Odd number				
Multiple of 4				
Factor of 120				

- Think of other numbers to add to each section.

3C Properties of numbers

Explore

- Look at the pattern in the first line. Fill in the missing parts:

$2^2 - 1^2 = 4 - 1$ = $\qquad$ $3^2 - 2^2 = 9 - 4$ = $\qquad$
$4^2 - 3^2 =$ $\qquad$ = $\qquad$ $5^2 - 4^2 =$ $\qquad$ = $\qquad$

- Now continue the pattern for all the square numbers to 10^2:

6^2 7^2
8^2 9^2
10^2

- Write a sentence to describe the pattern of your answers:

The answer is the same as _____

- Do the same for this pattern:

$3^2 - 1^2 = 9 - 1$ = $\qquad$ $4^2 - 2^2 =$ $\qquad$ = $\qquad$
$5^2 - 3^2 =$ $\qquad$ = $\qquad$

- Write a sentence to describe the pattern of your answers:

The answer is double _____

- What do you think will happen in this pattern?

$4^2 - 1^2 = 16 - 1$ = $\qquad$ $5^2 - 2^2 =$ $\qquad$ = $\qquad$
$6^2 - 3^2 =$ $\qquad$ = $\qquad$

- Were you correct? Test your idea.

3D Counting on and back

Discover

- Use a set of digit cards.
- Work with a partner.
- You and your partner pick three digit cards and each make a 3-digit number.
- Add your numbers together.

- Keep one of the numbers whole and add the hundreds, then tens and then ones of the other number.
- Next take the smallest number away from the largest in the same way.
- Record your work in this table. Show how you added and subtracted.

Numbers made	Addition	Subtraction

Do you think this is a useful mental calculation strategy?
_____
_____

- Explain your thinking:

   I think it is useful because _____
   _____
   _____

   I don't think it is useful because _____
   _____
   _____

3D Counting on and back

Explore

- Use a set of digit cards.
- Pick four cards.
- Make a 4-digit number.
- Write your number here: _____
- Rearrange the cards to make another 4-digit number.
- My second number is: _____

- Make up a problem that involves either adding or subtracting your two numbers.
- Solve your problem.
- Do this several times.
- Make up new problems by picking another set of four digit cards.
- Record your work in this table.

Problem	Show how you found the answer

Mental Calculation Strategies

3E Near multiples

Discover

- Write six 2-digit numbers.
  Your numbers must end with 1 or 2.
- Write your numbers here:

- Now add each number to 157 using the near multiple strategy you used during the lesson:

- Explain what you did:

  First I _____

  _____

  Then I _____

  _____

  Next I _____

  _____

- Use this strategy to solve these problems:

1. Alfie bought a DVD player for $236 and a DVD box set for $32.

   How much did he spend in total?

   - Explain how you found your answer:

48

2. Chevi paid $648 for a flight to London.
   He also paid $81 for a night in a hotel.
   How much did he spend in total?
   - Explain how you found your answer:

3. Amira had $249.
   Her mother gave her another $42.
   How much did she have in total?
   - Explain how you found your answer:

3E Near multiples

Explore

- Answer these calculations. Use the strategy you practised during the lesson.

1. 2004 − 1995
2. 3010 − 2999
3. 2003 − 1992
4. 3004 − 2997
5. 5001 − 4996
6. 4005 − 3996
7. 1012 − 999
8. 6008 − 5990
9. 5010 − 4900
10. 6009 − 5980

- Now solve these problems. Show how you work them out.

1. There were 906 adults and children at the carnival.

   799 were children.

   How many adults were at the carnival?

2. There were 2012 people at the football match.

   1980 were adults.

   How many were children?

3. Polly wants to buy a pair of jeans.

   The jeans cost $40.

   Polly has $35.90.

   How much more money does she need?

3F Which strategy?

Discover

- Work with a partner.
- Use two dice.
- Take it in turns to throw the dice, six times in total.
- Write each number down as you throw it.
- Use your numbers to make two 3-digit numbers.
- With your partner make up a problem using your numbers. Your problem must involve **addition**.
- Now solve your problem!
- Show the strategy you used.

For example:

Dice numbers thrown: 5, 3, 3, 6, 2, 4

Numbers made: 643 and 532

Problem:

Freddie and Sal washed some plates.

Freddie washed 643 plates.

Sal washed 532 plates.

How many plates did they wash altogether?

Solution: 643 + 500 + 30 + 2 = 1175.

They washed 1175 plates altogether.

Now it is your turn:

Problem 1

Dice thrown:

Numbers made:

Problem:

Solution:

Problem 2

Dice thrown:

Numbers made:

Problem:

Solution:

- Make up some more problems on a separate piece of paper.

Mental Calculation Strategies

3F Which strategy?

Explore

- Work with a partner.
- Use a set of digit cards.
- Take it in turns to pick six cards.
- Write each number.
- Use your numbers to make two 3-digit numbers.
- Next make up a problem using these numbers with your partner.

Your problem must involve **subtraction**. You can write or draw your problem.

- Now solve your problem!
- Show the strategy you use.

For example:

Cards picked: 8, 5, 3, 2, 9, 4.

Numbers made: 853 and 942.

Problem: Auzuma and Sam made some pancakes.

Auzuma made 942 pancakes. Sam made 853 pancakes.

How many more pancakes did Auzuma make than Sam?

Solution: $942 - 800 - 50 - 3 = 89$.

Auzuma made 89 more pancakes than Sam.

Now it is your turn:

Problem 1

Cards picked:

Numbers made:

Problem:

Solution:

Problem 2

Cards picked:

Numbers made:

Problem:

Solution:

- Make up some more problems on a separate piece of paper.

3G Multiplication strategies

Discover

- Work with a partner.
- Use some counters of two different colours.
- Decide which colour counter you will each use.
- Take it in turns to pick two numbers to multiply together.
- Write down your answer.
- Then cover the numbers you used with your counters. You cannot use these numbers again.
- Continue until you have used all the numbers.
- Your total score is the sum of the answers to all your multiplications.
- Find your total score.
- The winner is the player with the highest total.

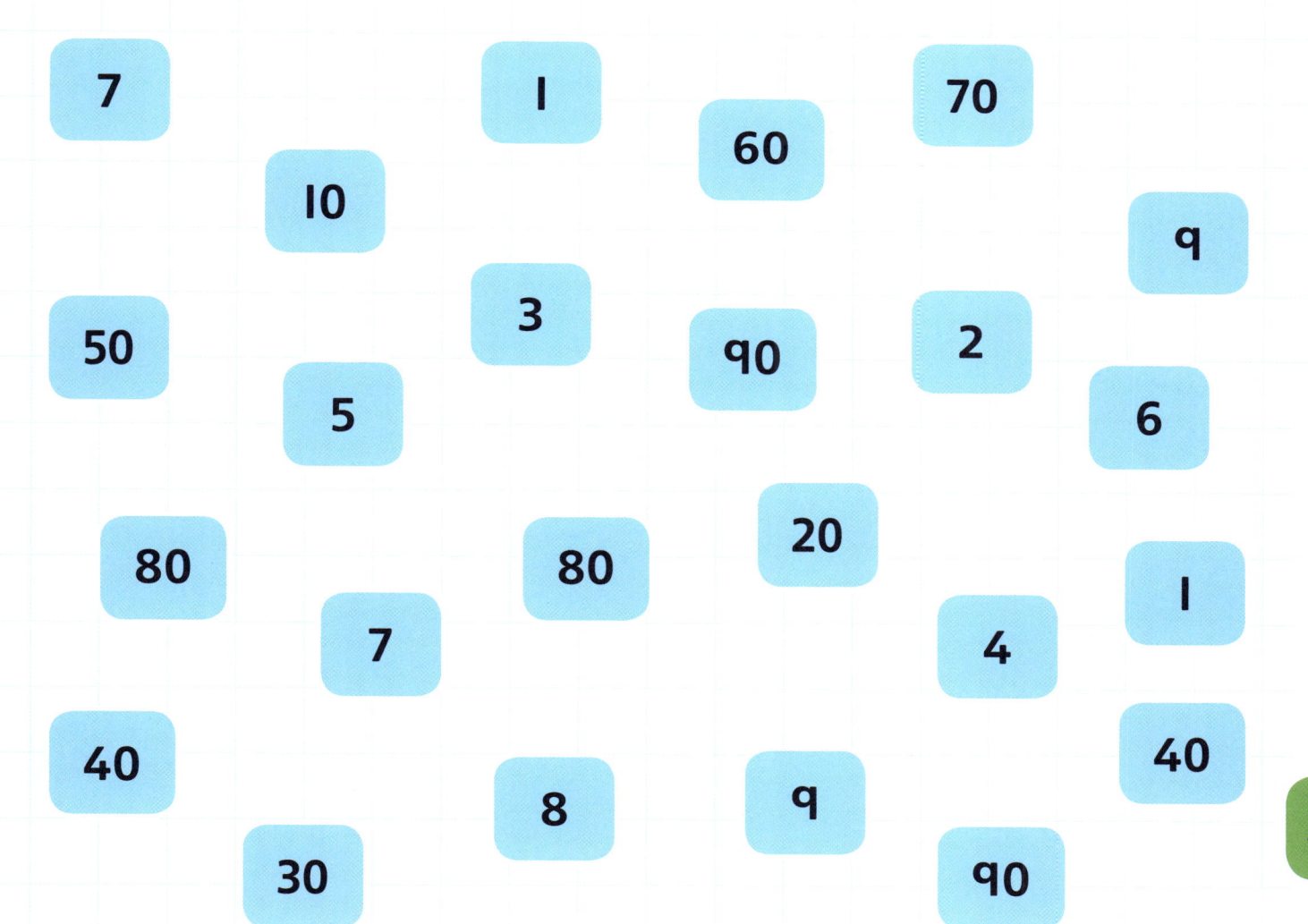

7   1   70
   10   60
      3   9
50   90   2
   5   6
         20
80   80   1
   7   4
40   40
      8   9
   30   90

Who won?

3G Multiplication strategies

Explore

- Work with a partner.
- Use two sets of digit cards.
- Shuffle the digit cards.
- Place the cards face down on the table.
- Take it in turns to pick a digit card.
- Multiply that number either by 19 or 21.
- Put the answer on the ladder.
- Put the numbers on the ladder in order with the smallest numbers at the bottom.
- Write numbers that can't fit onto the ladder beside the bin.

The player with the most numbers on the ladder wins.

3H More multiplication strategies

Discover

- Use two sets of even-numbered digit cards.
- Use your digit cards to make six different 2-digit numbers.
- Multiply your numbers by 25.
- Use the strategy you practised in the lesson.

For example:

Number made: 28

Multiply by 25: 28 × 100 = 2800.

Half of 2800 = 1400,

Half of 1400 = 700,

so 28 × 25 = 700

Now it's your turn!

Number made:
Multiply by 25:

Number made:
Multiply by 25:

Number made:
Multiply by 25:

Number made:
Multiply by 25:

Number made:
Multiply by 25:

Number made:
Multiply by 25:

3H More multiplication strategies

Explore

- Practise multiplying numbers by:
  - 4 by multiplying by 2 and doubling
  - 6 by multiplying by 3 and doubling
- Complete the table to show that you can do this. An example is shown:

Number	Multiply by 4 by multiplying by 2 and doubling	Multiply by 6 by multiplying by 3 and doubling
9	$9 \times 2 = 18$, $18 \times 2 = 36$ Answer = 36	$9 \times 3 = 27$, $27 \times 2 = 54$ Answer = 54
8		
6		
12		
14		
21		
15		
25		
32		

3I Doubling and halving

Discover

- Match each number with its double.
  For example:
  3.6 is double 1.8

2.3	18.2
7.5	13.4
1.8	6.8
4.6	17.8
9.1	12.4
3.4	3.6
5.2	18.8
6.7	15
8.9	4.6
2.7	10.4
6.2	9.2
9.4	5.4

- Match each number with its half.
  For example:
  8.2 is half of 16.4

14.6	4.2
13.2	2.8
12.6	6.3
9.4	8.2
8.4	7.9
7.2	4.9
5.6	8.5
16.4	7.3
15.8	4.7
11.6	6.6
9.8	3.6
17	5.8

- Write a sentence to explain how you found the **doubles and halves**:
  _____

- Choose ten numbers from the above lists. Write them in **ascending** order:
  _____

Mental Calculation Strategies

57

3I Doubling and halving

Explore

Can you double and halve multiples of 10 and 100?
- Complete the table to show how well you can do this.
- Show how you worked out your answers.

Number	Double it!	Halve it!
240		
360		
780		
540		
490		
170		
160		
640		
1400		
2300		
7600		
8700		
4900		
5200		
3500		
6800		

3 Mental calculation strategies

Connect

- Work in a group.
- Work out the answers to these calculations in as many different ways as you can.
- Use the mental calculation strategies from the lesson.
- Use a piece of A3 paper to share your ideas. Which strategies were the best for each calculation?
- Write them in the spaces:

Do you all agree?

2009 − 1992

3649 + 2990

4567 − 1234

2781 − 1217

320 × 25

132 × 6

12 400 + 12 500

3 Mental calculation strategies

Review

- Work with a partner.
- Think of all the mental calculation strategies from this Unit.
- Write them around this mind map.

Some key words are given as a reminder.

- Give examples to show what you know.

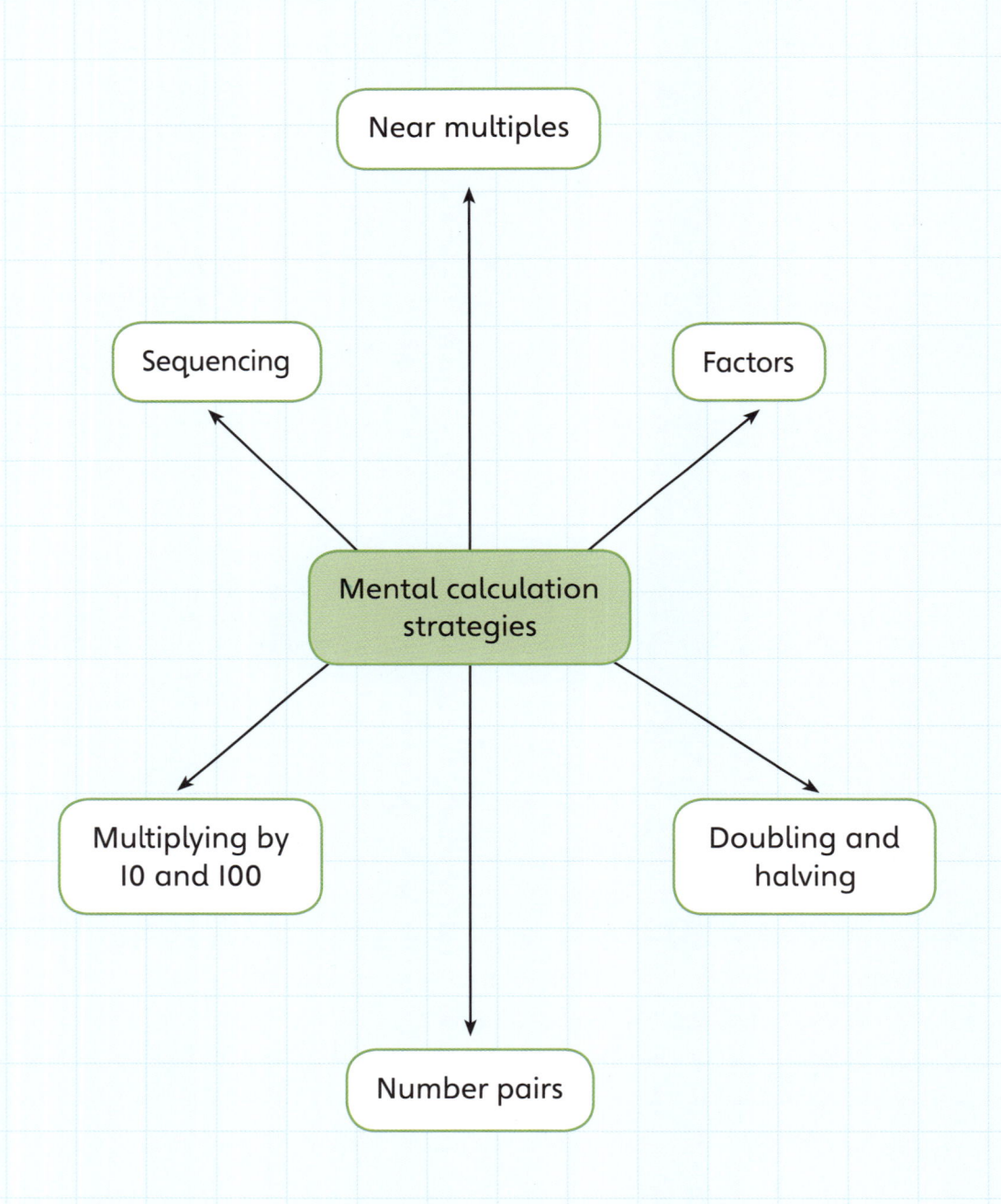

```
         Near multiples
              ↑
Sequencing         Factors
       ↖        ↗
     Mental calculation
         strategies
       ↙    ↓    ↘
Multiplying by        Doubling and
  10 and 100            halving
              ↓
         Number pairs
```

60

4 Written Calculation

Engage

How can you answer these calculations?
- Talk with your group. Then answer the calculations!

1. 5689 + 2387

2. 4678 + 3595

3. 5683 − 2385

4. 4098 − 2743

5. 458 × 7

6. 784 × 6

7. 364 ÷ 7

8. 586 ÷ 9

4A Addition and subtraction

Discover

- Pick four digit cards. Make a 4-digit number.
- Repeat this so that you have two 4-digit numbers.
- Find the total of your two numbers. Use your favourite strategy.
- Check using another strategy.
- Now find the difference of the two numbers. Use your favourite strategy.
- Check using another strategy.
- Use this grid to record your work:

Numbers made	Find the total	Check	Find the difference	Check

4A Addition and subtraction

Explore

Waleed wants to buy a football game with his pocket money.

He has $20.

He checks the price of the game at different shops.

How much will he have left if he buys the game from each shop?

- Use the space in the table to show your working:

Shop	Price	Change from $20 (show your working)
Games Galore	$10.50	
Fun and Games	$9.45	
Football Mania	$11.15	
Footie Stuff	$8.93	
Team Work	$12.01	
Games for You	$10.79	

He wants to buy two games.

Has he got enough money?

Which shop can he go to?

4B More addition and subtraction

Discover

- Pick four cards. Make a 4-digit number.
- Write your number down.
- Rearrange the cards to make another 4-digit number.
- Write an addition or subtraction problem with your two numbers.

> **For example:** My digit cards are   1, 3, 7, 8
>
> My two numbers are   3871 and 1783
>
> My problem is   Andy and Penny collect stamps. Andy has 3871 stamps. Penny has 1783 stamps. How many more stamps than Penny does Andy have?

- Solve your problem.
- Repeat this several times.
- Write both addition and subtraction problems.
- Record your work in this table:

Digit cards	4-digit numbers	Problem	Show how you found the answer

4B More addition and subtraction

Explore

- Pick three digit cards. Make a 3-digit number.
- Rearrange the digits so that you have four different 3-digit numbers.
- Find the total of your four numbers.
- Write the strategy you used in the table.
- Now find the differences between pairs of your numbers.
- Write the strategy you used in the table.

Digits picked	Numbers made	Addition	Difference	Strategy

- Write an addition and subtraction problem for one of your sets of numbers:

4C Multiplication

Discover

- Pick three digit cards. Make a 3-digit number.
- Pick another card.
- Multiply your 3-digit number by the fourth number you picked.
- Choose which strategy to use.
- Record your work in this table:

Numbers picked	Multiplication calculation	Strategy used	Answer

4C Multiplication

Explore

- Put 2-digit numbers into the problems in the table.
- Use digit cards to make your numbers.
- Write your numbers in the spaces and solve the problems.

Problem	Workings	Answer
Shar collects stamps in a stamp album. His album has <u>36</u> pages. On each page there are <u>98</u> stamps. How many stamps does he have altogether?	36 × 98 | | 90 | 8 | 30 | 2700 | 240 | 2940 6 | 540 | 48 | 588	2940 + 588 = 3528
Bethany has _____ bags of shells. Each bag has _____ shells inside it. How many shells does Bethany have altogether?		
Thor packs boxes of biscuits. In each box there are _____ packets. In total he packs _____ boxes. How many packets does he need?		

Written Calculation

Problem	Workings	Answer
The store receives _____ crates of cola. In each crate there are _____ cans of cola. How many cans are there altogether?		
Sam delivers _____ trunks of books to a library. In each trunk there are _____ books. How many books does he deliver?		
Write a similar problem about footballs. Solve it.		
Write a similar problem about pineapples. Solve it.		
Write another similar problem.		

4D Multiplying decimals

Discover

- Pick two digit cards.
- Make a digit number with one decimal place.
  For example: I pick a 3 and a 6 and make 3.6.
- Pick a third digit card.
- Multiply your decimal number by the number on the third card.
- Decide which method to use to multiply them together.
- Repeat this six times.
- Show your working in this table:

Numbers picked	Calculation	Solution

4D Multiplying decimals

Explore

- Work with a partner.
- Take it in turns to roll the dice.
- Each roll, move that number of squares.
- Choose a single-digit number.
- Multiply the number on the square by this number.
- Ask your partner to check your answer:

  For a correct answer, stay on that square.

  For an incorrect answer, miss your next turn.
- Decide which method to use and write it in the square.
- The first person to arrive at the end wins.

Start	× 2.7				
	× 1.6	× 9.7	× 4.7	× 8.5	× 9.2
					× 4.3
	× 8.4	× 2.6	× 5.4	× 3.6	× 9.3
	× 7.3				
	× 2.6	× 3.1	× 5.8	× 7.7	End

4E Division

Discover

- Take it in turns to pick three digit cards and make a number.
- Throw the dice. Divide your number by the number on the dice.
- You can score points during each turn. Your remainder, if you have one, is the number of points you score.

> For example: I pick a 2, a 3 and an 8
>
> I make 283
>
> I throw a 5
>
> I divide 283 by 5
>
> The answer is 56 remainder 3
>
> I score 3 points

- Record your work in this table:

Player 1				Player 2			
Cards picked	Number thrown	Calculation	Points scored	Cards picked	Number thrown	Calculation	Points scored
		Total score				Total score	

Who won?

4E Division

Explore

Work with a partner.

- Make up some grouping problems.
- Give your problems to your partner to answer.

> **For example:** Wally has 96 DVDs.
>
> He stores them in his DVD cabinet.
>
> Each shelf in the cabinet holds 8 DVDs.
>
> How many shelves does he fill?

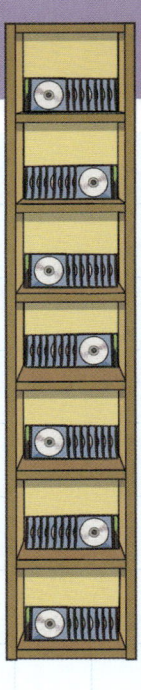

- Write your problems here:

Problem 1

Problem 2

Problem 3

Problem 4

Problem 5

4F More division

Discover

- Answer each problem with your partner.
- Check your answers using multiplication.

1. Bethan and Riley collect 240 pebbles from the beach.

   They share them equally into four piles.

   How many pebbles are there in each pile?

   What fraction of the pebbles have you found?

2. Sanjit buys 56 cakes.

   He shares them equally onto eight plates.

   How many cakes are there on each plate?

   What fraction of the cakes have you found?

3. Ahmed has $48.50.

   He puts the money equally into two bags.

   How much money is in each bag?

   What fraction of the money have you found?

4. Antonio buys 35 chocolate bars.

   He shares them between five people.

   How many chocolate bars does each person get?

   What fraction of the chocolate bars have you found?

5. Shelly organises 138 beads into six piles.

   How many beads are in each pile?

   What fraction of the beads have you found?

4F More division

Explore

- Answer these problems:

Do you need to round the answer up or down?

1. Ashley puts tins of beans in boxes.

   He has 125 tins.

   There are 8 tins in each box.

   How many boxes does he need?

   Did you round your answer up or down? Why?

2. Beverly has a box of 120 books.

   She puts them on shelves.

   She puts seven books on each shelf.

   How many shelves does she use?

   Did you round your answer up or down? Why?

3. Manni has 146 potatoes.

   He puts them in bags of eight.

   How many bags does he fill?

   Did you round your answer up or down? Why?

4. Rachel makes bracelets.

   She has 218 beads.

   She puts 9 beads on each bracelet.

   How many bracelets can she make?

   Did you round your answer up or down? Why?

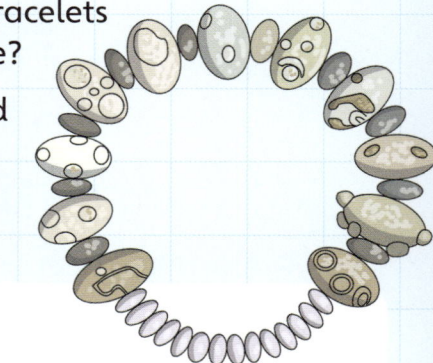

4G Brackets

Discover

- Make up a calculation that involves two operations.
- Work out the answer, working from left to right.
- Add brackets so that you get a different answer.
- Work out the different answer.

> **For example:** I use the operations ÷ and ×
>
> I write the calculation 36 ÷ 9 × 2
>
> The answer working from left to right is:
>
> 36 ÷ 9 = 4, 4 × 2 = 8
>
> I add brackets to give 36 ÷ (9 × 2)
>
> The answer is now 36 ÷ 18 = 2

- Try to include different operations in your calculations.
- Record your work in this table:

Calculation	1st solution	With brackets	2nd solution

Use this space for notes:

Use this space for notes:

Calculation	1st solution	With brackets	2nd solution

Use this space for notes:

Use this space for notes:

Use this space for notes:

4G Brackets

Explore

- Work with a partner. Talk about what you are doing.
  The factors of 12 are: 1, 12, 2, 6, 3, 4
- List the factors of these numbers:

  15

  16

  20

  48

  64

You can use factors to answer multiplication calculations.

> **For example:** $48 \times 12 = 48 \times 2 \times 2 \times 3$
> $= 96 \times 2 \times 3$
> $= 192 \times 3$
> $= 300 + 270 + 6$
> $= 576$

- Use factors to answer these multiplication calculations:

  20×16

  48×15

  64×12

4 Written calculation

Connect

- In your group make up some word problems for each of these calculations.
- Use a piece of A3 paper to share ideas.
- Solve your problems. Choose a method to use for each.
- Check your answers using the inverse operation.

For example: 7454 − 6398

Our problem: There were 7454 supporters at a football match. There were 6398 supporters at the next football match. How many more supporters were there at the first match than at the second match?

Answer: 6398 + 2 + 600 + 454 = 7454, 600 + 454 + 2 = 1056

4563 + 2766

Our problem:

Answer:

739 × 6

Our problem:

Answer:

267 ÷ 7

Our problem:

Answer:

5693 + 3976

Our problem:

Answer:

376 ÷ 4

Our problem:

Answer:

4 Written calculation

Review

With your partner, try to remember the addition, subtraction, multiplication and division facts from this Unit.

- Write them around symbols on this mind map.
- Write the vocabulary you know too.
- Give examples to show what you know.

For example: + is the addition sign

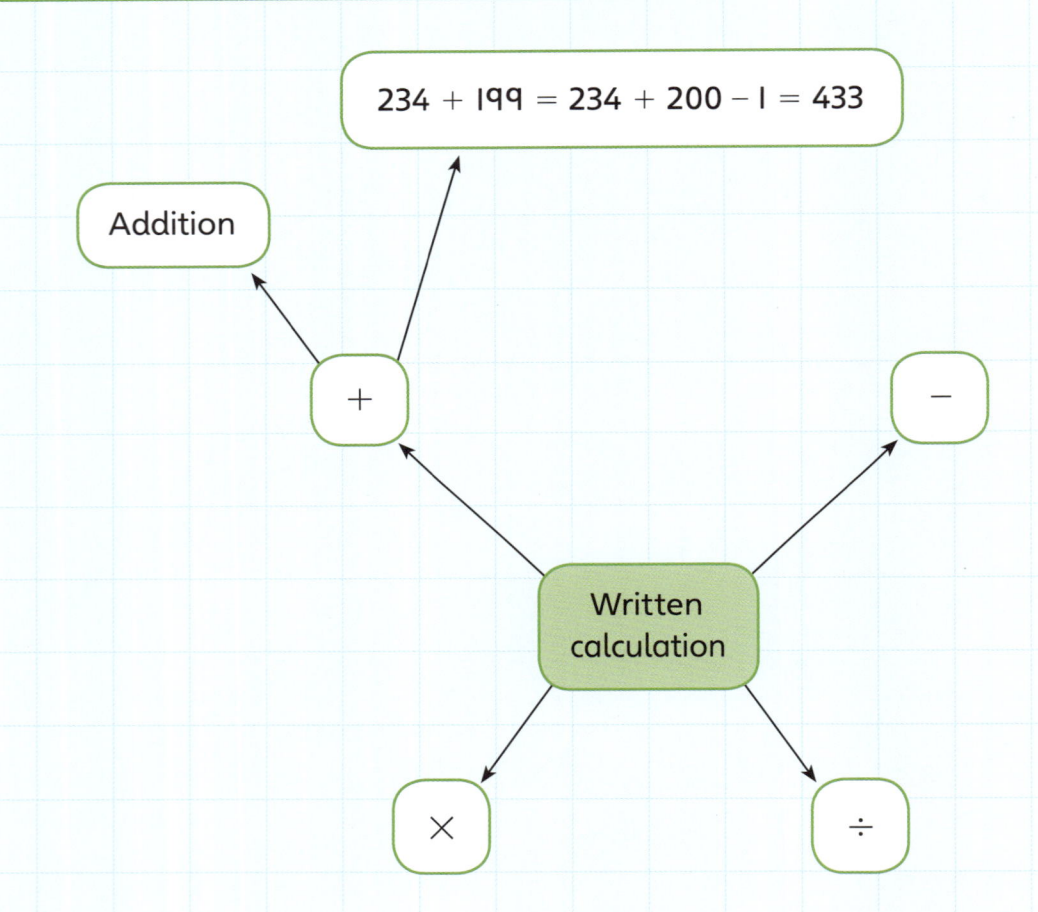

$$234 + 199 = 234 + 200 - 1 = 433$$

$$234 + 199 = 234 + 200 - 1 = 433$$

Addition

+

−

×

÷

Written calculation

5 Shape

Engage

cube

cuboid

cylinder

cone

hemisphere

tetrahedron

pyramid

triangular prism

sphere

- In your group take it in turns to describe one of the shapes in the picture. (Do not say the name of the shape.)
- The rest of your group needs to guess which shape you are describing.
- Make sure your descriptions include:

  The number and shapes of the **faces**.

  The number of **edges**.

  The number of **vertices**.

  Is the shape a **prism** or not?
- Next, draw lines to match each shape with its name.

5A Triangles

Discover

- Name the **triangle** to match each clue.
- Then draw the triangle as accurately as you can.

Use a ruler to draw your triangles.

Clues	Name	Drawing
All angles acute, all sides different lengths		
All angles the same		
One right angle, two sides the same length		
One right angle, all sides different lengths		
Two equal sides and two equal angles		

5A Triangles

Explore

- Draw these triangles.
- Then describe the triangles.

Include the types of angle (acute, obtuse, right) and the number of lines of symmetry.

You can use the descriptions on the Discover page to help you.

1. An isosceles triangle with 3 acute angles

   Describe your triangle

2. An isosceles triangle with a right angle

   Describe your triangle

3. An equilateral triangle

   Describe your triangle

4. A scalene triangle with a right angle

   Describe your triangle

5. A scalene triangle with an obtuse angle

   Describe your triangle

5B Symmetry

Discover

- Work with a partner.
- Use a piece of A4 paper.
- Fold it in half (**horizontally** or **vertically**).
- Use the fold as a mirror line.
- Decide who will go first.
- Draw and colour a small 2D shape on one side of the mirror line.
- Your partner should match it on the other side of the mirror line.
- Continue to do this until you have created a pattern.
- Draw your pattern here:

- Do this again on another piece of paper.
- This time fold the paper **diagonally**.
- Draw your pattern here:

5B Symmetry

Explore

- Cut a square out of a piece of paper.
- Put a dot in one corner.
- **Rotate** the square through a quarter turn.
- Repeat this 4 times until the dot is in the same corner as at the start.

This means a square has order of **rotational symmetry** 4.

Trace these shapes and rotate them to see if they have rotational symmetry:

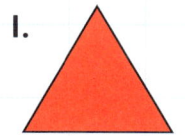

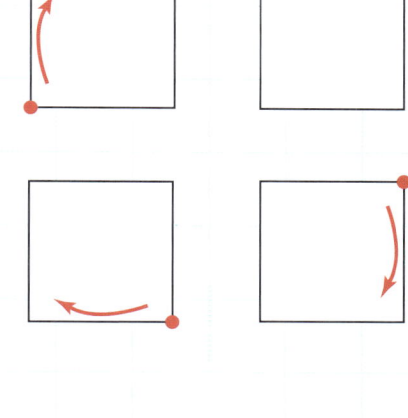

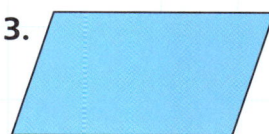

1.
2.
3.

Draw one example of each shape with rotational symmetry.

Polygon	Sketch	Order of rotational symmetry	Order of reflective symmetry
Regular quadrilateral			
Regular hexagon			
Regular octagon			
Regular pentagon			

Draw another example of a polygon with rotational symmetry.

5C 3D shape

Discover

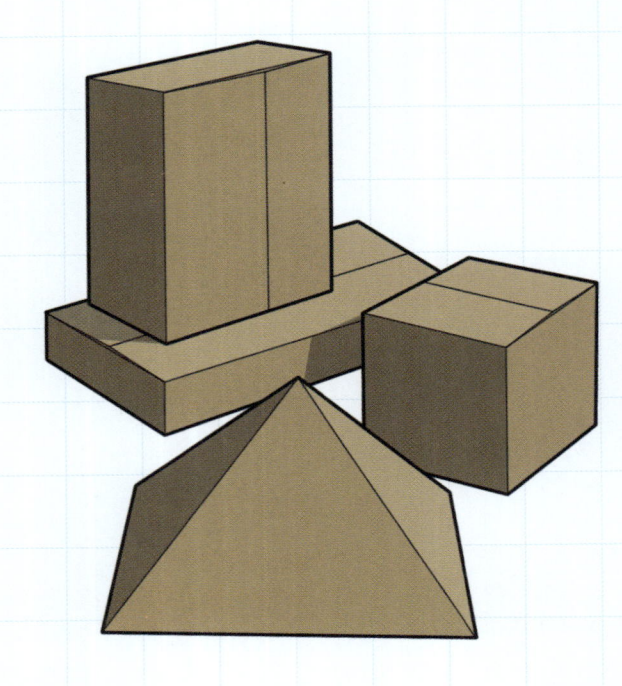

If you made a mistake tell your partner what your mistake was.

You have made the net of a **pyramid**.

Before you open your box sketch the **net** here.

Use what you have learned to sketch the net for this **hexagonal pyramid**.

Now open your box and sketch the net here. Where you correct?

5C 3D shape

Explore

Here are the faces of some 3D shapes.

- Under them, write all the 3D shapes that you think have this shape as a face. For example: A cuboid has a rectangle as a face.

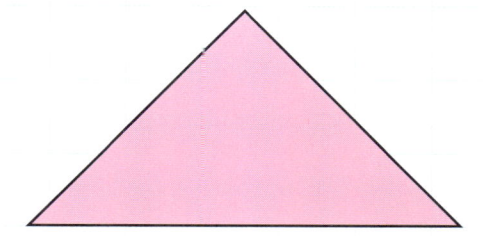

Now try to remember the nets you have made for 3D shapes.

- Draw the nets of:
  - A cube
  - A square-based pyramid
  - A cuboid
  - A triangular prism

5D Lines

Discover

- Make up a pattern.

In your pattern include 6 pairs of **parallel** lines and 6 sets of **perpendicular** lines.

Here is an example of the start of a pattern:

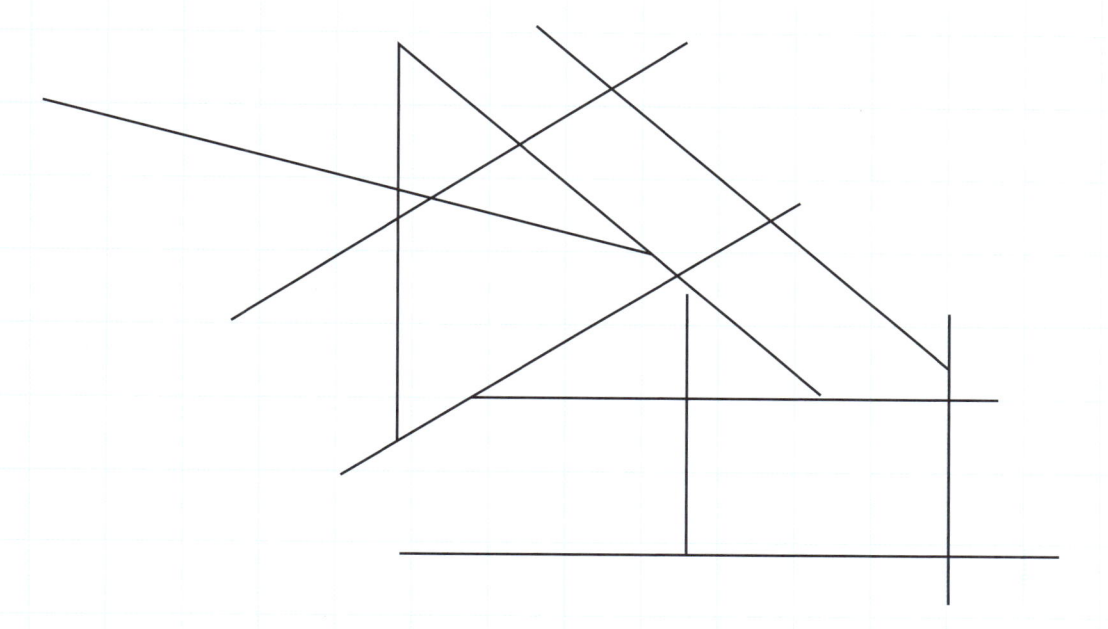

- Draw your pattern in this space:

What shapes can you see in your pattern?

- List them here:

- Colour your shapes.
- Choose a different colour for each shape.

5D Lines

Explore

Benji painted this picture.

How many pairs of parallel lines can you see?

How many sets of perpendicular lines can you see?

5E Angles

Discover

- In the rectangles below, draw at least two more examples of right, acute and obtuse angles.
- Use a ruler when you draw the lines.

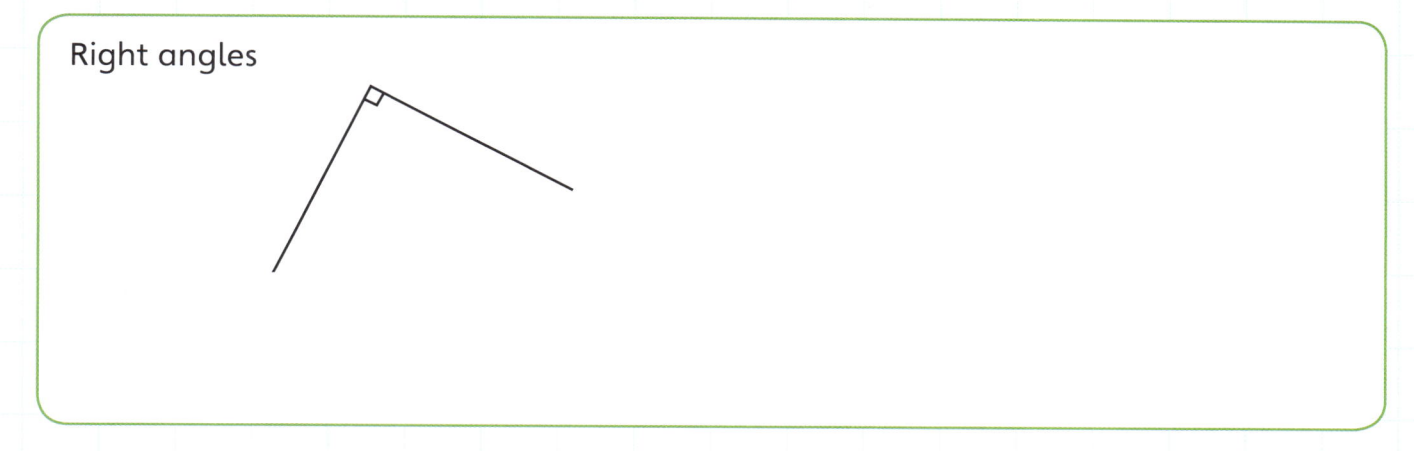

Right angles

Acute angles

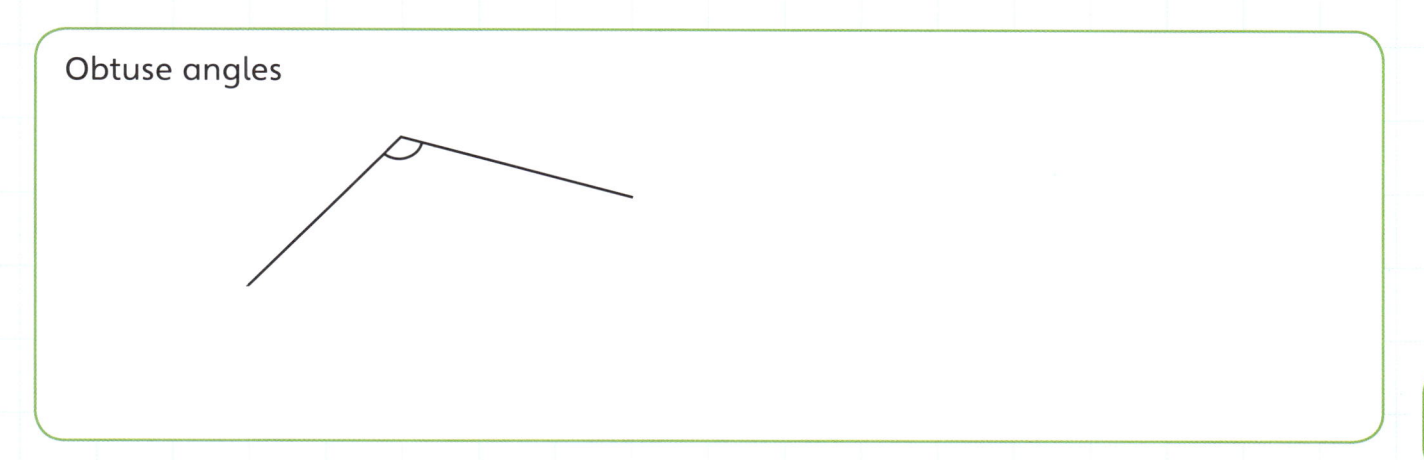

Obtuse angles

- Now draw three shapes that have at least one right angle, one acute angle and one obtuse angle inside them.
- Name each shape.

For example:

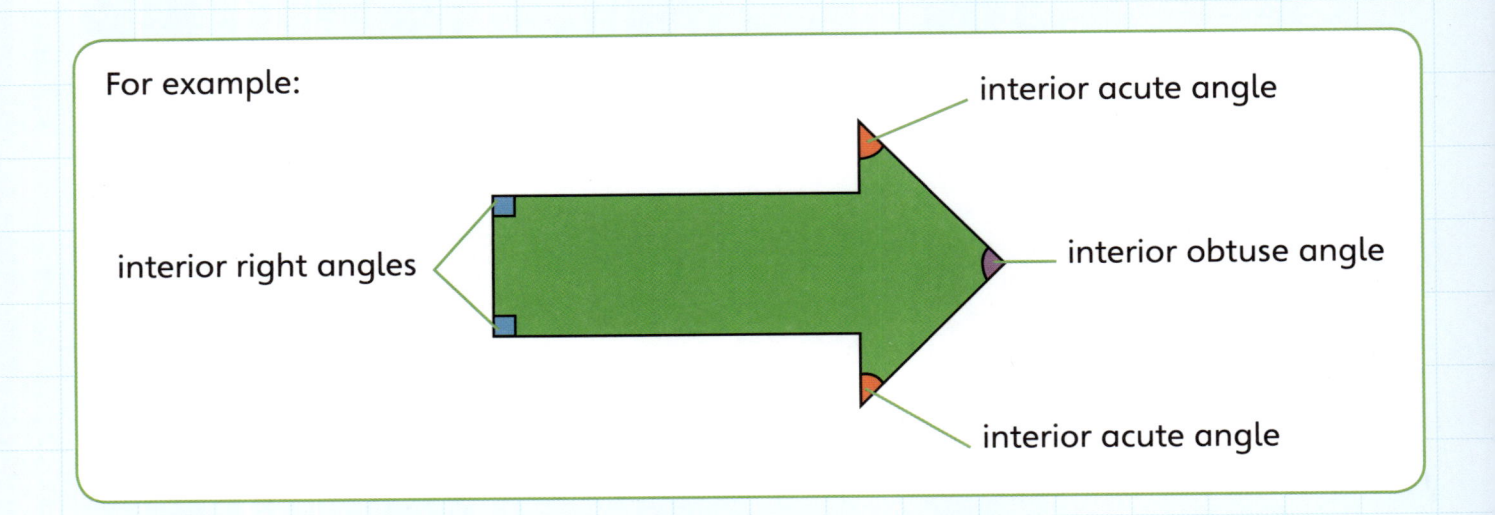

interior acute angle

interior right angles

interior obtuse angle

interior acute angle

5E Angles

Explore

Use your knowledge of angles.

- Estimate these angles:

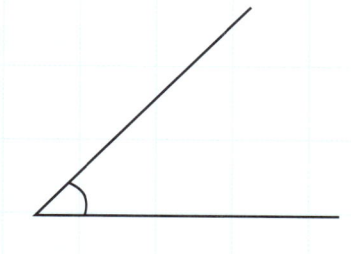

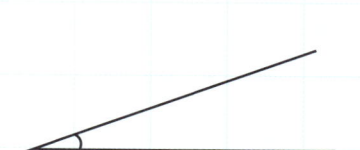

- Now measure the angles to the nearest 5°.
- Write your estimate and the actual size beside the angles.
- How close were your estimates?

- Measure the angles inside these triangles:
- Label each angle.
- Find the total of the angles in each triangle.

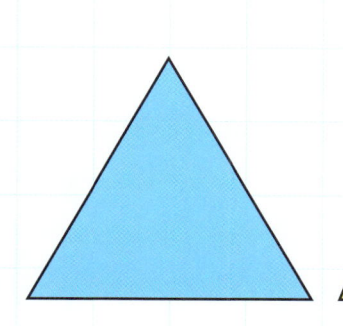

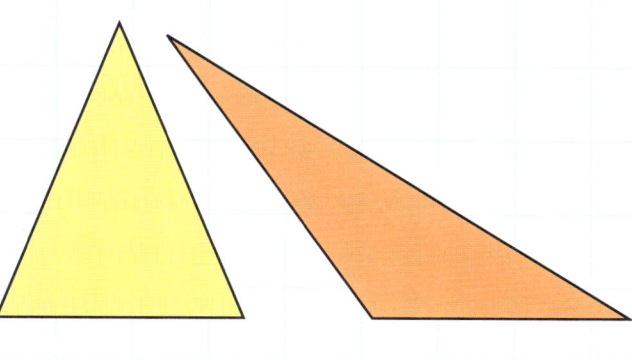

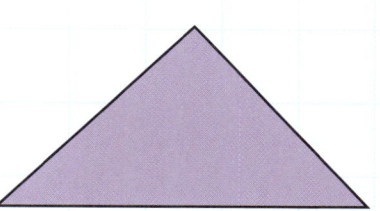

Total of the angles:

Tammy thinks that the total of the angles inside a triangle is 180°.
Is she correct?

5F Polygons

Discover

- Try out these investigations.
- Do one at a time.
- Compare your results with your partner.

Investigation 1

- Draw a rectangle on a piece of squared paper.
- Draw an imaginary diagonal mirror line.

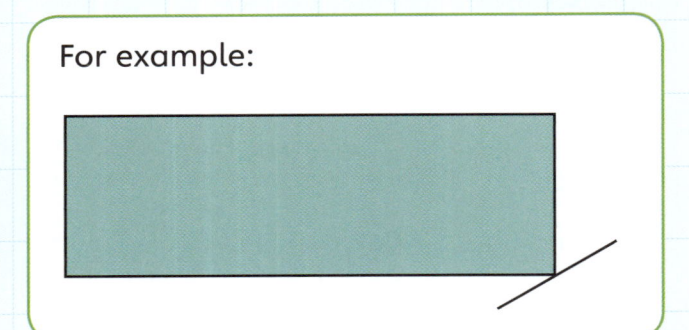

For example:

- Now draw where you predict the reflection will be:

- Check your prediction with a mirror.

Experiment with other regular and irregular polygons (for example: triangle, square and hexagon).

Investigation 2

- Fold a piece of squared paper in half and then in half again.
- Make a pattern in one of the quarters.
- Pass your pattern to your partner.
- Choose one of the other quarters and copy the pattern so that it is symmetrical.
- Do this twice more so that the squared paper is a complete pattern.
- Draw the pattern you made here:

5F Polygons

Explore

- Write the name of each shape:
- Now draw all the lines of symmetry on each shape.

Name:
Number of lines of symmetry:

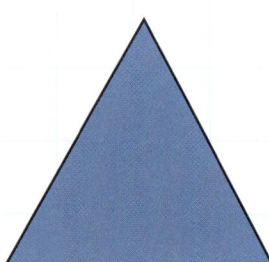

Name:
Number of lines of symmetry:

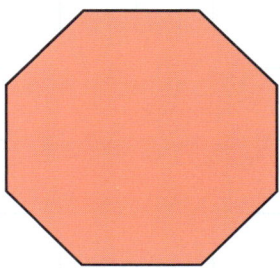

Name:
Number of lines of symmetry:

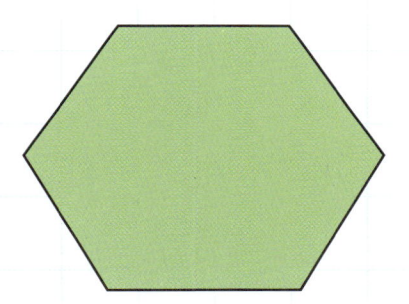

Name:
Number of lines of symmetry:

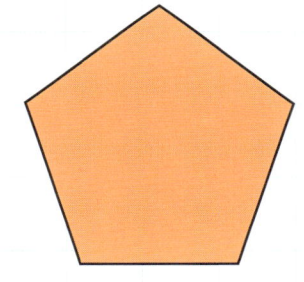

Name:
Number of lines of symmetry:

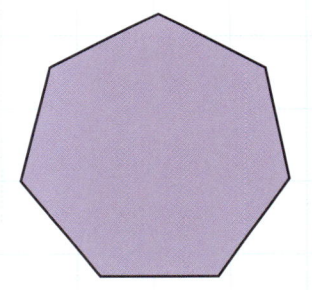

Name:
Number of lines of symmetry:

What do you notice about the number of lines of symmetry in a regular shape?

Investigate!

- Trace over the shapes twice on a piece of plain/tracing paper.
- Cut out one of each shape.
- Place the cut-out shape on the other drawing.
- Find out how many times you can rotate the shapes until you are back at the beginning.

What do you notice about the number of times you can rotate a regular shape?

5 Shape

Connect

- Work in a group of four.
- Use some modelling clay to make two of these shapes.
- Make sure that your group makes all of the listed shapes.
  - Sphere
  - Cylinder
  - Cone
  - Cube
  - Cuboid
  - Pyramid
  - Triangular prism
  - Tetrahedron
- Describe the shapes you made here:

Name of shape:

Number of faces:

Number of curved surfaces:

Shapes of the faces:

Number of vertices:

Number of edges:

Where you see these shapes in real life:

Name of shape:

Number of faces:

Number of curved surfaces:

Shapes of the faces:

Number of vertices:

Number of edges:

Where you see these shapes in real life:

- Together, look at your descriptions of all eight shapes.

Together, focus on the cube, tetrahedron and pyramid.

- Draw a net for each of them here:

5 Shape

Review

- With your partner, estimate these angles:

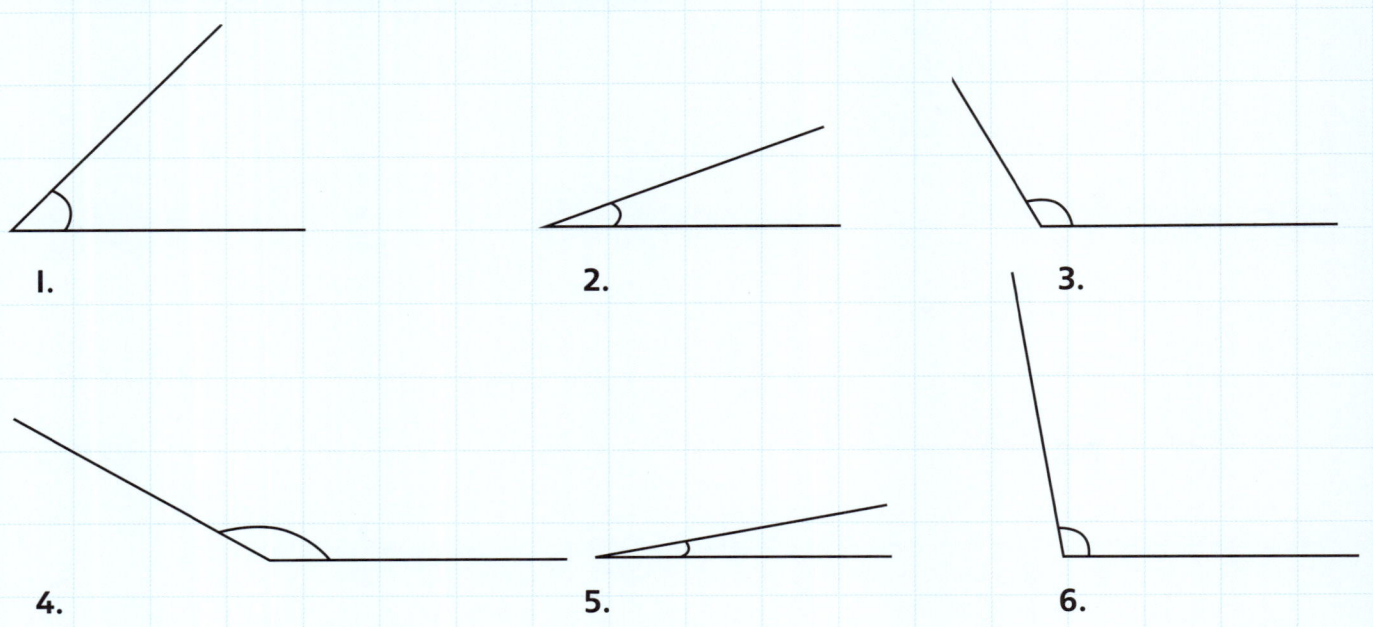

1.

2.

3.

4.

5.

6.

- Now measure the angles.
- Work out the difference between your estimate and the actual measurement.

The student with the estimate closest to the actual measurement scores a point.

The winner is the student with the most total points.

	Estimate	Actual	Who was closest?
1			
2			
3			
4			
5			
6			

Who won?

6 Position and Movement

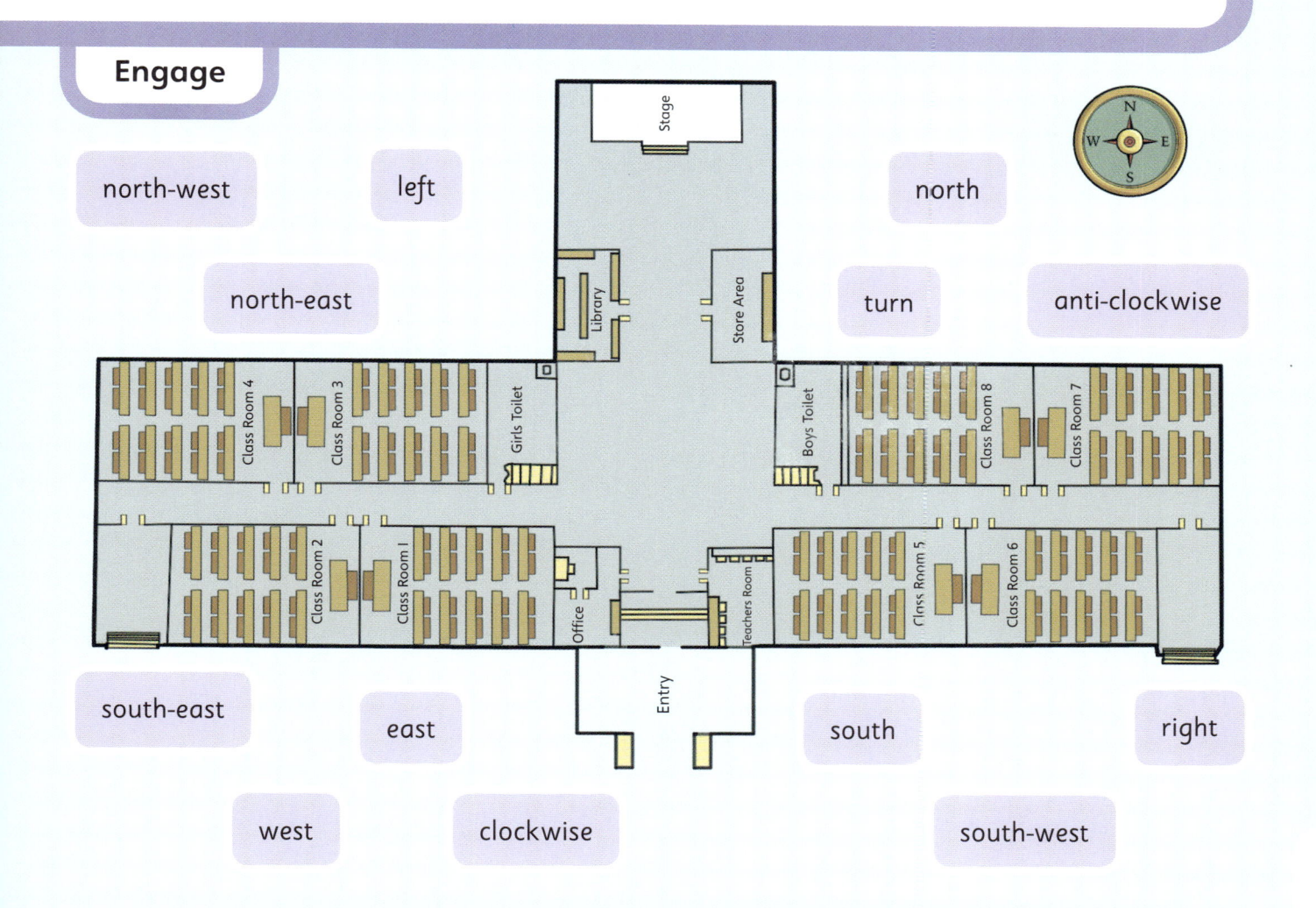

Engage

north-west | left | north

north-east | turn | anti-clockwise

south-east | east | south | right

west | clockwise | south-west

- Take it in turns to give instructions to go from the entry hall to different classrooms.
- Write your set of instructions on another sheet of paper.
- Draw your route on the classroom plan.
- Complete these sentences:

  The office is north of _____.

  The office is east of _____.

  Classroom 2 is in a _____ direction from the office.

- Together make up some more sentences like this.

6A Coordinates

Discover

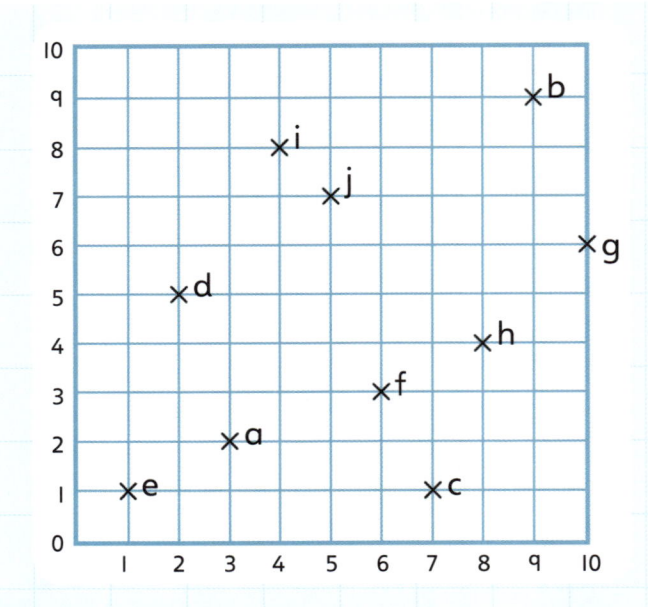

The coordinates of cross **b** are (9, 9).

- Write the coordinates of the other crosses:

a   b (9, 9)   c   d   e

f   g   h   i   j

- Now plot these coordinates on this grid: (10, 10), (6, 7), (2, 7), (4, 6), (7, 3), (9, 5). Mark each with a cross.

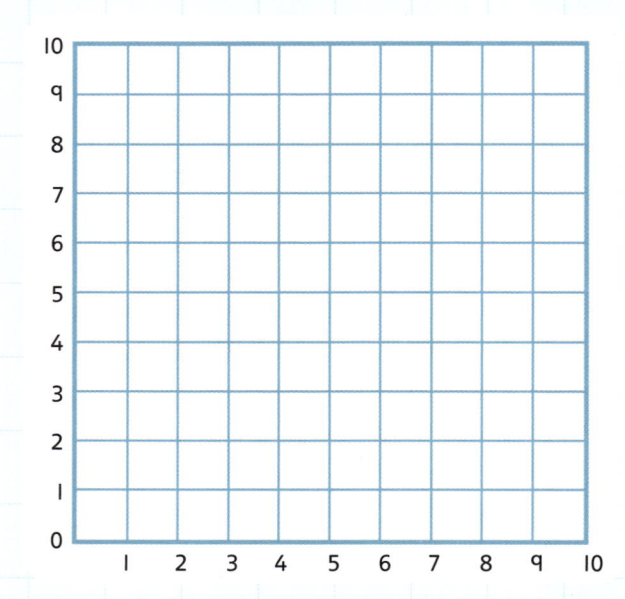

- Join the crosses.

What is the shape?

6A Coordinates

Explore

- Work with a partner.

  Here is a coordinate grid:

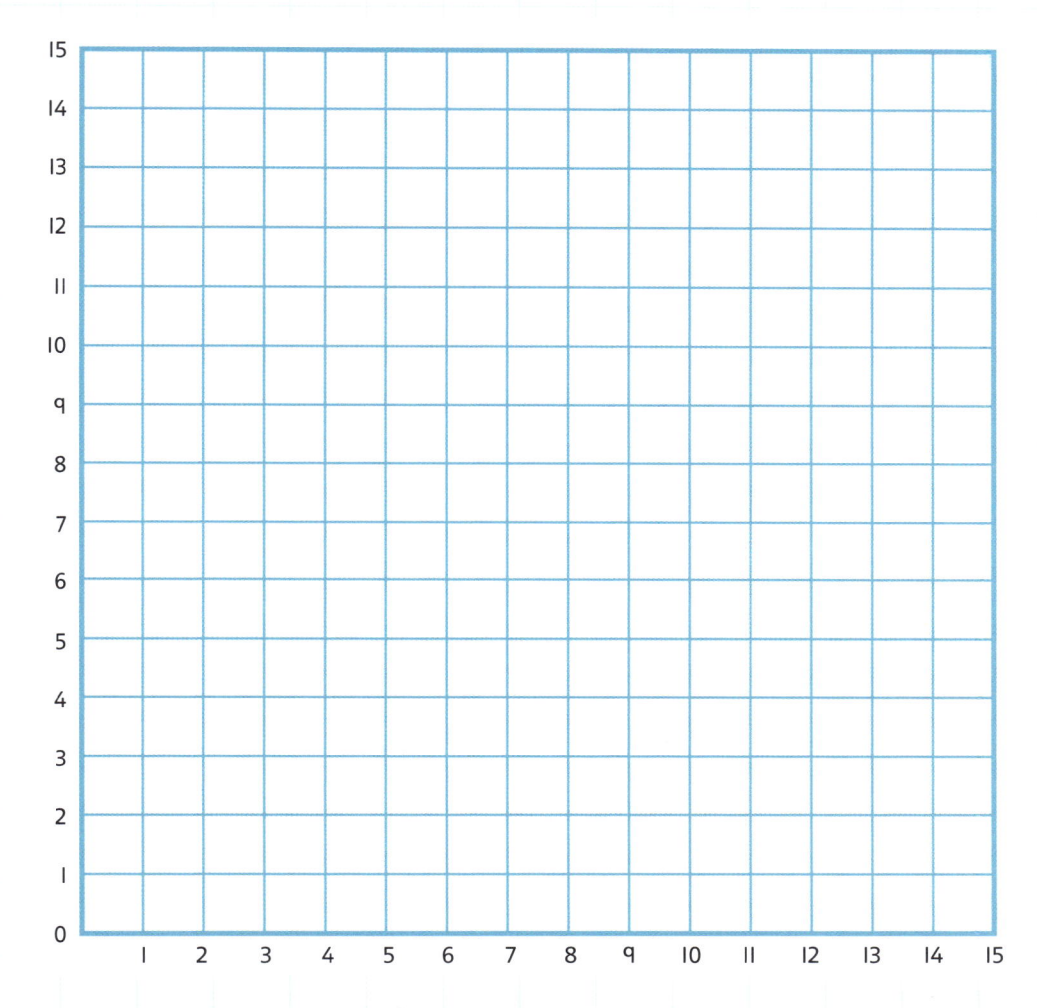

- Plot three corners of a small square onto the grid.
- Give your grid to your partner.

  Your partner works out where the other coordinate goes to make the square.

- Record the coordinates of the whole square.
- Now plot two corners of a scalene triangle.

  Your partner then draws the third corner and records the coordinates.

  Ask your partner: *Why did you place the third coordinate here?*

  I placed the third coordinate here because I think _____

  _____

- Repeat this for:

  a) a triangle with a right angle

  b) a triangle with an obtuse angle.

6B Reflection

Discover

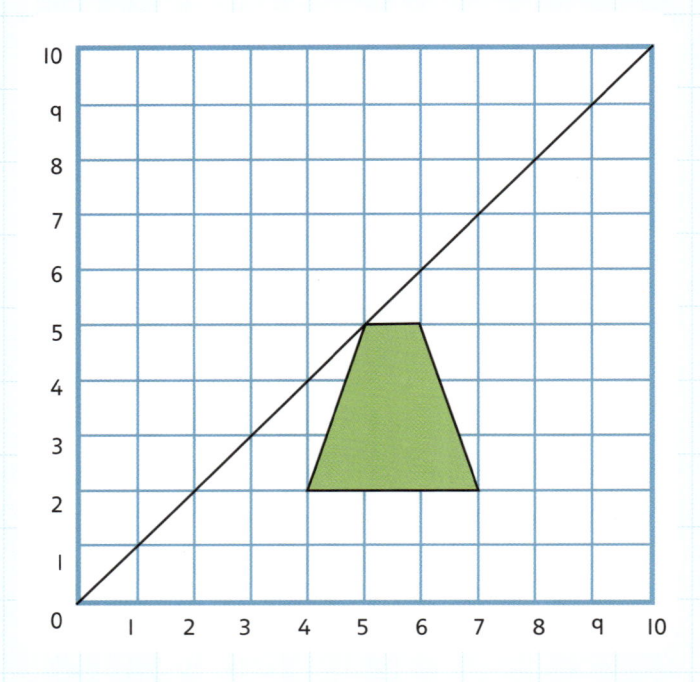

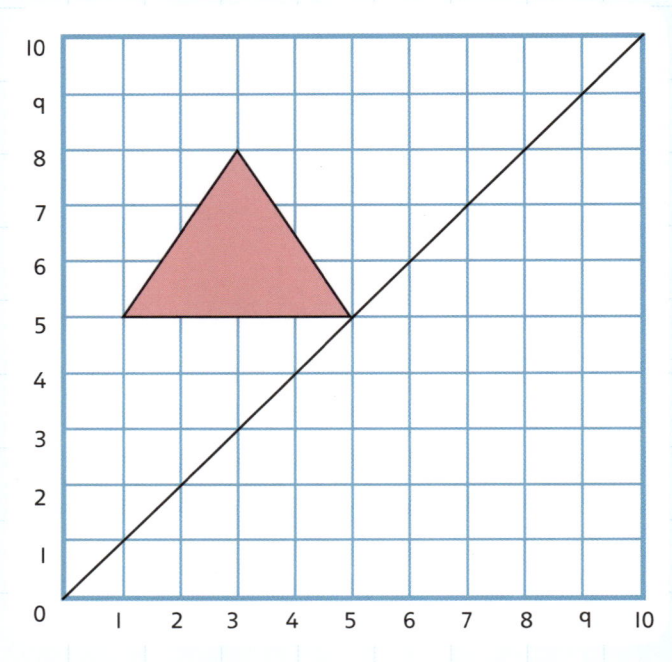

- Look closely at the shapes on the grids.
- Reflect each shape across the diagonal mirror line.

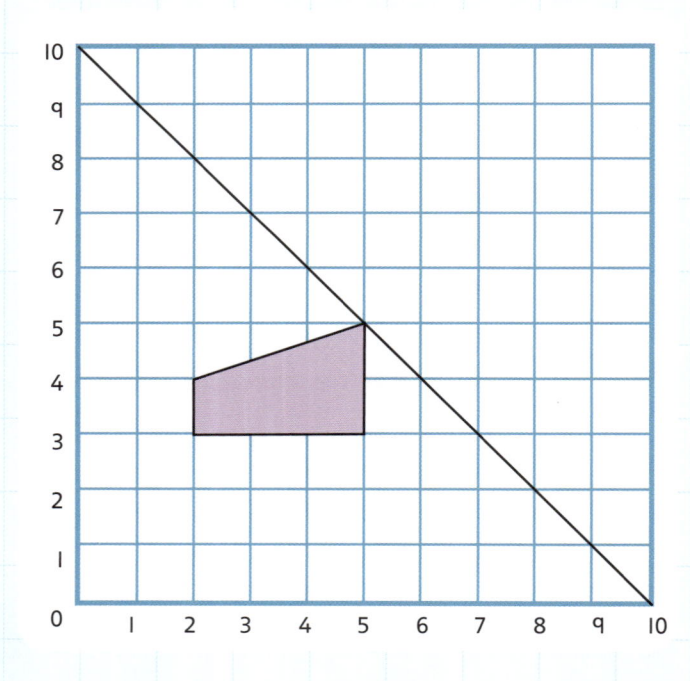

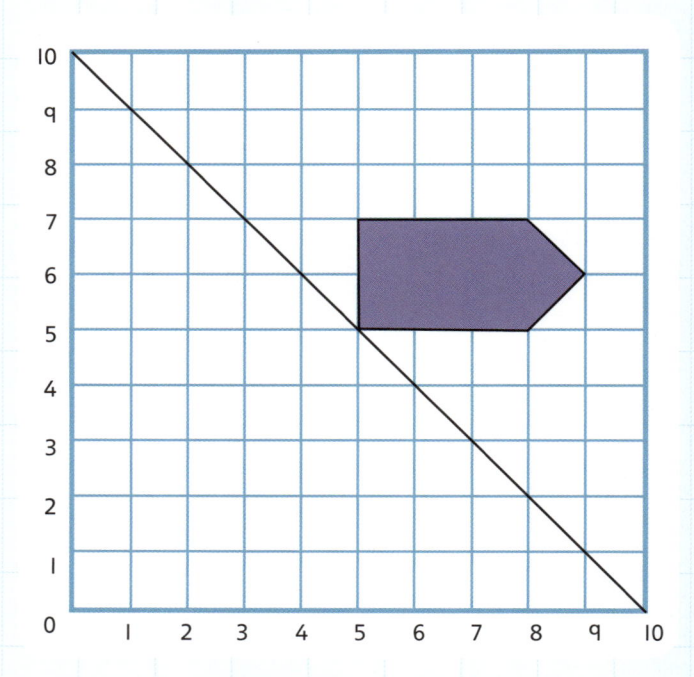

6B Reflection

Explore

- Look closely at the shapes on the grids.
- Reflect the shapes in each axis so that you have four shapes in each grid.

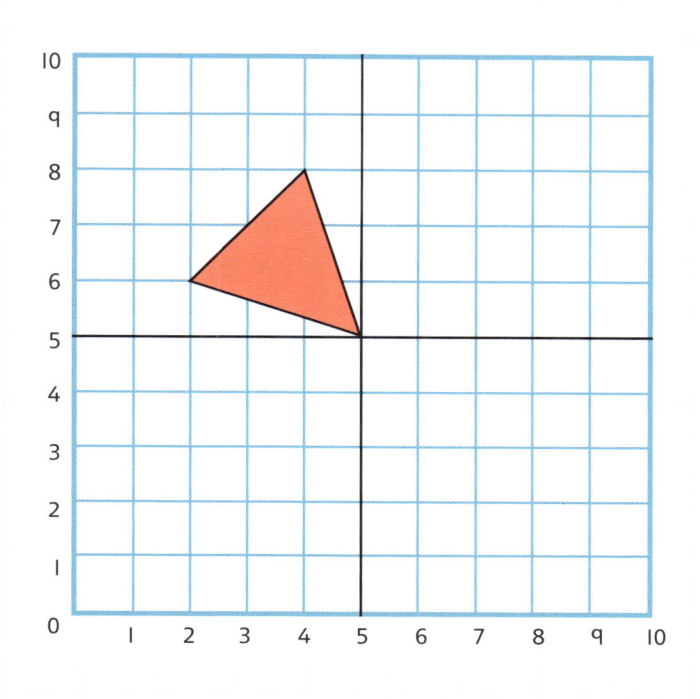

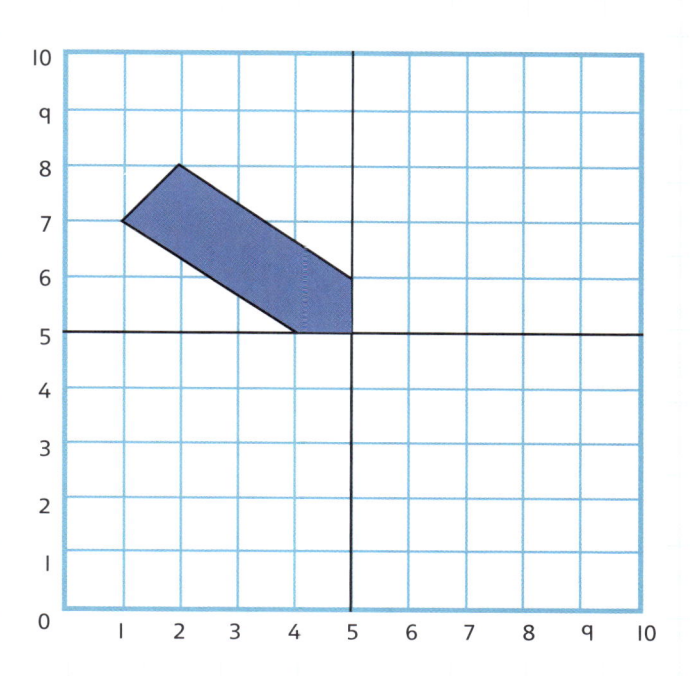

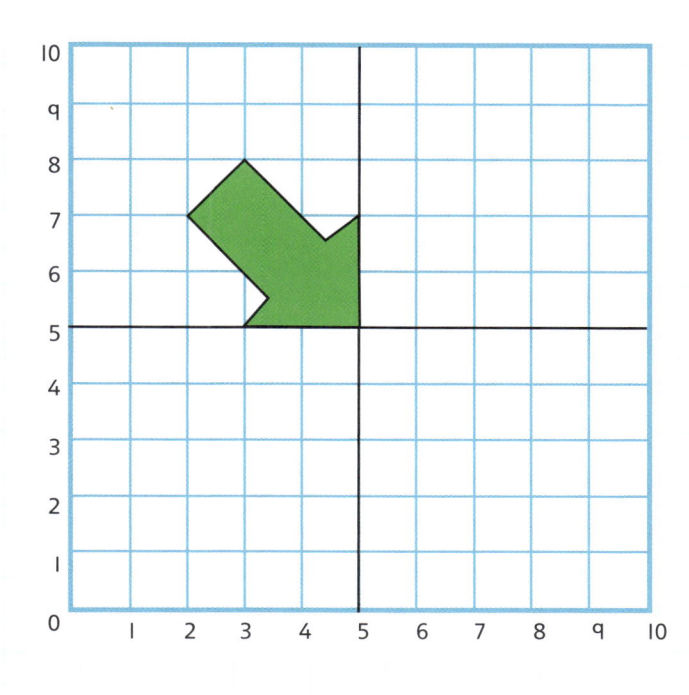

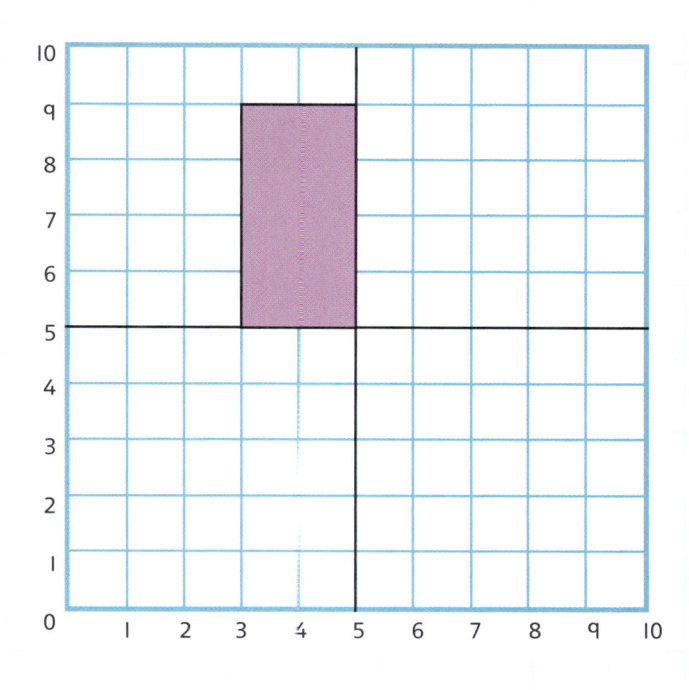

6C Translations

Discover

- Draw a symbol (such as a star or a circle) in the top left corner of this grid.
- Translate your shape following the instructions.
- After each move draw your symbol in its new position.

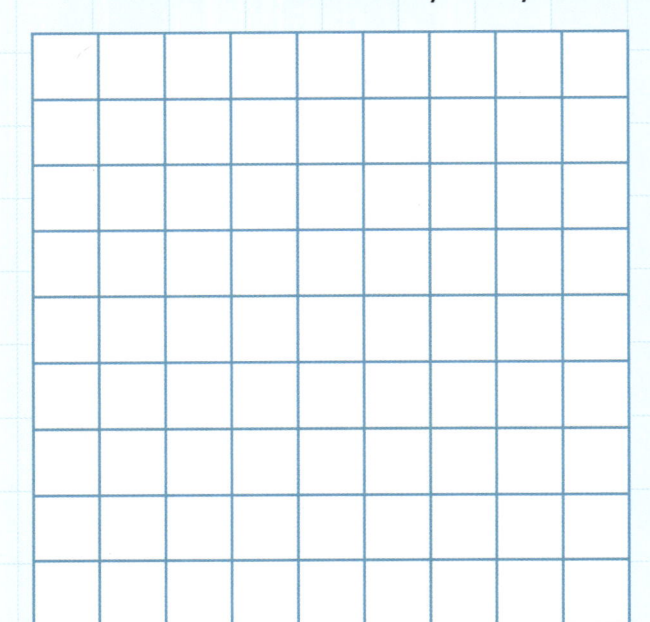

3 to the right

4 down

1 to the left

3 down

7 to the right

6 up

2 to the left

1 up

2 to the right

- Now write your own set of instructions and follow them on this grid:

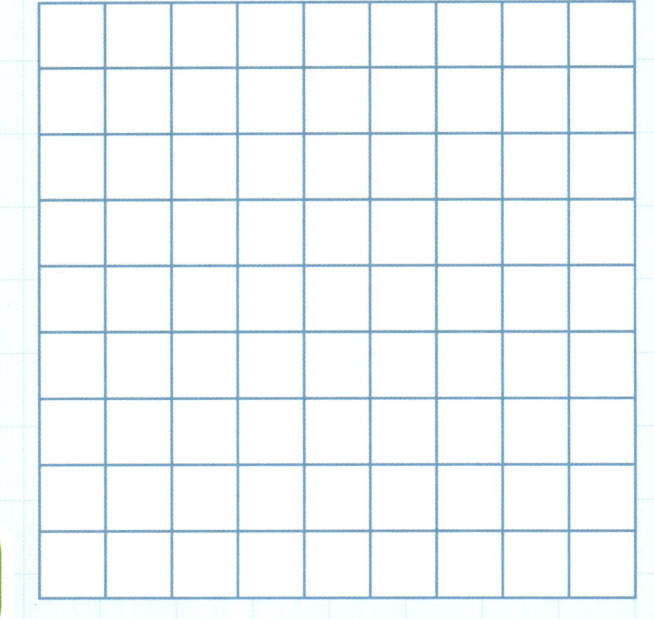

- Give your instructions to your partner to follow on a piece of squared paper.

Is their result the same as yours?

6C Translations

Explore

This grid shows five shapes translated from one area to another:

- Describe the translations to your partner.

Don't go into or across the shaded squares!

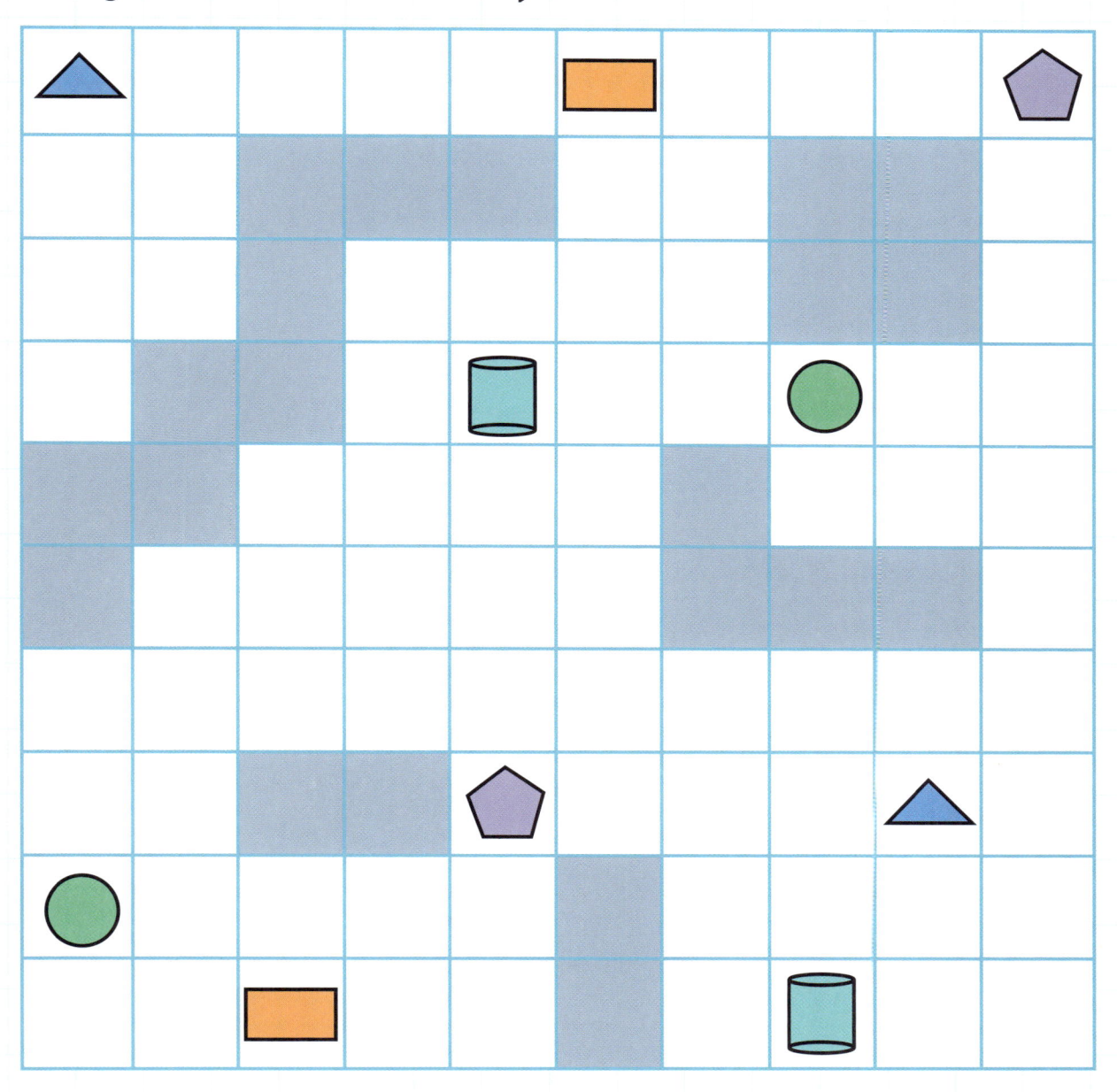

- Draw the translations on the grid.
- Now write your instructions here.

Use simple instructions. For example: use L for left, R for right, D for down.

My translations:

- Now add some more shapes or symbols to the grid and translate them.
- Give the grid to a partner.
- Ask your partner to write a possible translation for each shape.

6 Position and movement

Connect

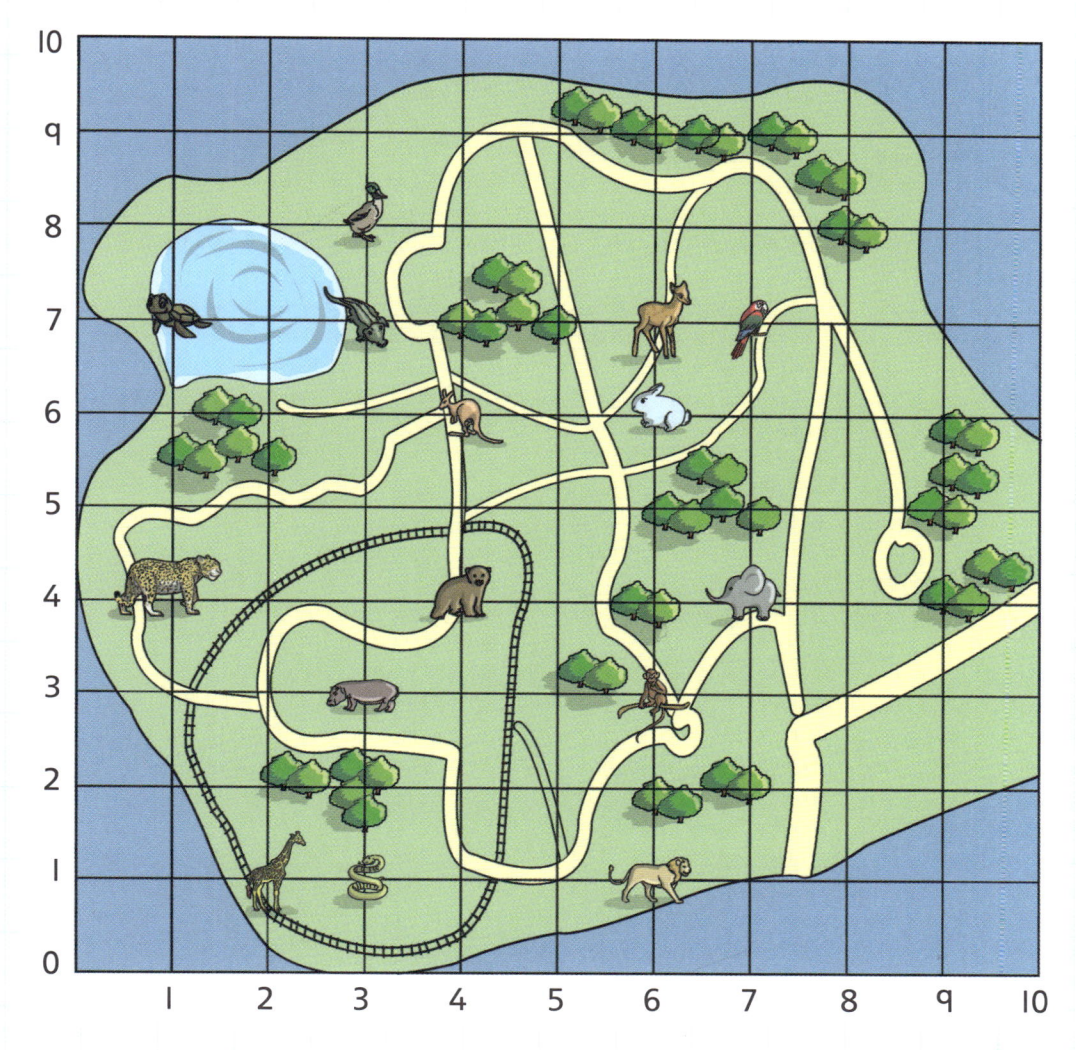

- Find the coordinates for the different animals.
- Give the coordinates of an animal to your group.

Ask: *What animal am I thinking of?*

Take it in turns to do this.

- Together make a key for the map. Include a picture of each animal and its name.

- Show your key here:

6 Position and movement

Review

- Work as a group.

Carol is doing a parachute jump.

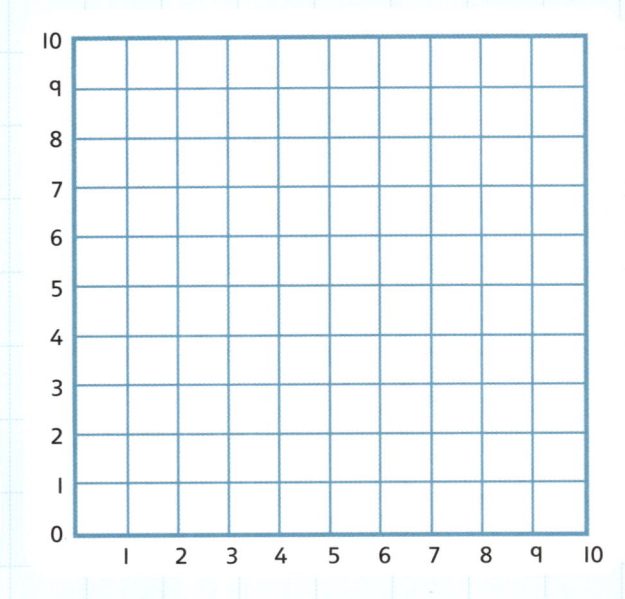

Michael is doing a parachute jump on another day when the wind is blowing quite strongly.

He has different map coordinates for the area he must land on.

The coordinates are different from Carol's coordinates:

- Carol's area is translated two squares to the right.
- It is then reflected vertically in $y = 6$.
- It is then rotated 90° in a clockwise direction from the bottom right-hand corner.
- Draw the new landing area.

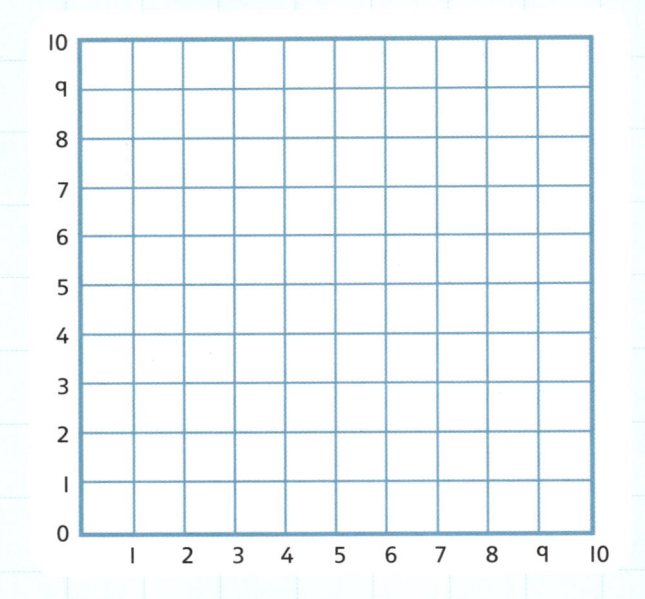

She has some map coordinates for the area she must land on. The coordinates are (2, 6), (2, 8), (4, 6) and (4, 8).

- Mark these coordinates on this grid:

7 Length, Mass and Capacity

Engage

- Draw lines to match each unit of **measure** with its abbreviation, equipment and picture.
- Use a different-coloured pencil for each unit of measure.

| Millilitre   |
| Gram |
| Metre |
| Kilogram |
| Centimetre |
| Kilometre |
| Litre |
| Millimetre |

| kg |
| mm |
| l |
| cm |
| km |
| m |
| ml |
| g |

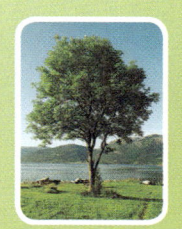

111

7A Units of measure

Discover

- Pick four digit cards.
- Make a 4-digit number.
- Your number is _____ **grams**.

1. Write your number in as many equivalent units as you can:

2. Round your grams to the nearest 10, 100 and 1000:

- Change the order of your digit cards to make a new number.
- Your number is _____ **millilitres**.

1. Write your number in as many equivalent units as you can:

2. Round your millilitres to the nearest 10, 100 and 1000:

- Change the order of your digit cards again to make a new number.
- Your number is _____ **metres**.

1. Write your number in as many equivalent units as you can:

2. Round your metres to the nearest 10, 100 and 1000:

- Change the order of your digit cards again to make a new number.
- Your number is _____ **millimetres**.

1. Write your number in as many equivalent units as you can:

2. Round your millimetres to the nearest 10, 100 and 1000:

7A Units of measure

Explore

Work with a partner.
- Match the equivalent units in the list at the foot of the page.
- Complete this table. An example is shown:

Mixed unit	Decimal unit	Smallest unit
1 kg 750 g	1.75 kg	1750 g

1 kg 750 g	1 l 224 ml	1 km 500 m	2550 g	1245 cm	10050 ml
1 m 10 cm	25 mm	6 kg 75 g	1.75 kg	1.224 l	12 m 45 cm
103 mm	1500 m	2.65 l	2 l 650 ml	2 cm 5 mm	1750 g
2.55 kg	6.075 kg	10 l 50 ml	6075 g	12 l 450 ml	2650 ml
12450 ml	10.05 l	12.45 m	12.45 l	1.5 km	110 cm
2.5 cm	10.3 cm	1.1 m	1224 ml	2 kg 550 g	10 cm 3 mm

Length, Mass and Capacity

7B Measuring length

Discover

Work in groups of four.

- Use modelling clay to make the longest 'snake' you can in 30 seconds.
- In your group, place your snakes in order, from shortest to longest.
- Now estimate the **length** of the shortest snake.
- Write your estimate in the table.
- Now use rulers or tape measures to measure the snake. Write the measurement in the table.
- Use this measurement to estimate the length of the second snake.
- Now measure the second snake.
- Repeat for the other two snakes in order.
- Record your estimates, measurements and differences in the table.
- Write your measurements in centimetres and in millimetres. For example: 124 mm, 12.4 cm.

Name	Estimate	Measurement	Difference

Did your estimating get better as you worked through the snakes?

7B Measuring length

Explore

Work with a partner.

- Choose two places on the map, one on the west side and one on the east.
- Find a route between the two places.
- Draw your route on the map.
- Use the scale (1 cm represents 2 km) to work out the distance between the two places.

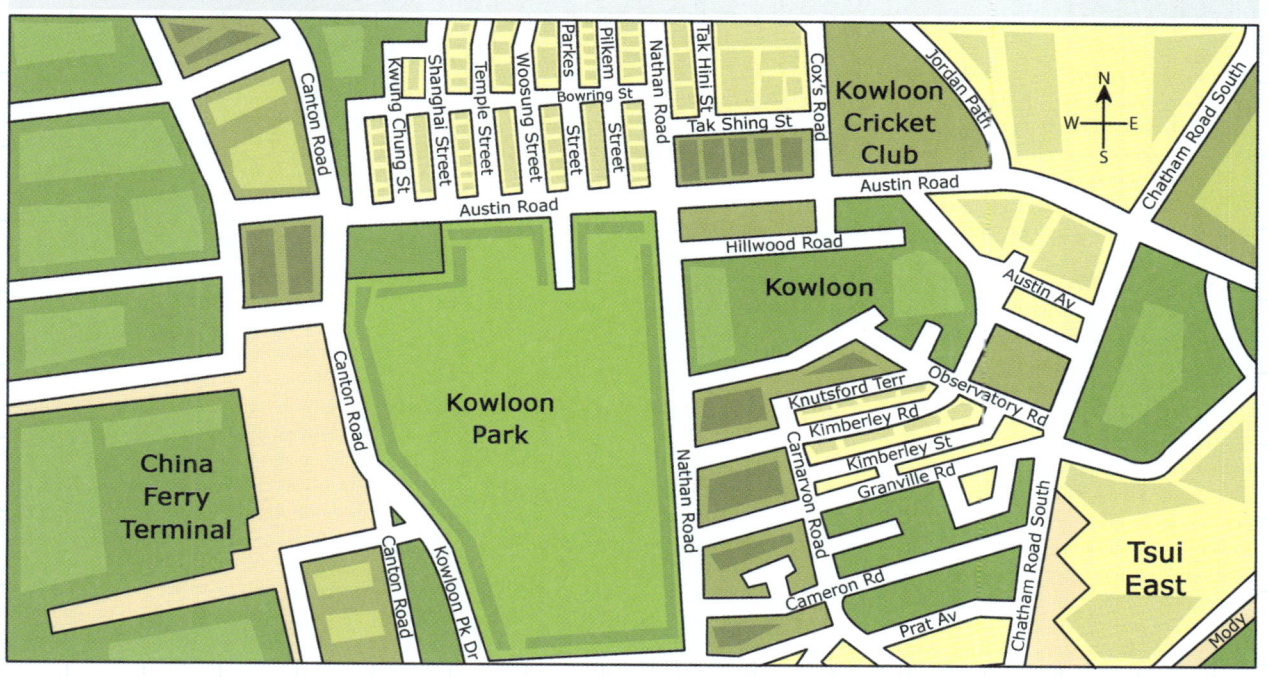

Scale 1 cm : 2 km

Which places did you choose?

I chose _____ and

_____

What is the distance between these two places?

The distance between them is _____

_____

- Now find two other routes between your two places – a longer route and a shorter route.
- Use different-coloured pencils to draw these routes on the map.

What are the two distances?

The longer distance is _____

The shorter distance is _____

7C Centimetres and millimetres

Discover

- Use a ruler to measure these lines.
- Write their lengths in three different ways: millimetres, centimetres and millimetres, and centimetres.

- Now draw lines of these lengths on a separate sheet of paper:

    1. 3 cm 4 mm
    2. 6.9 cm
    3. 85 mm
    4. 4 cm 2 mm
    5. 9.8 cm
    6. 16 mm

- Write these lengths in order in centimetres. Start with the shortest:

7C Centimetres and millimetres

Explore

Work with a partner.

Compare your measurements with your partner as you work.

- Use the width of your index finger to estimate the length of your hand. (The length of your hand is from your wrist to the end of your middle finger.)
- Write your estimate here:

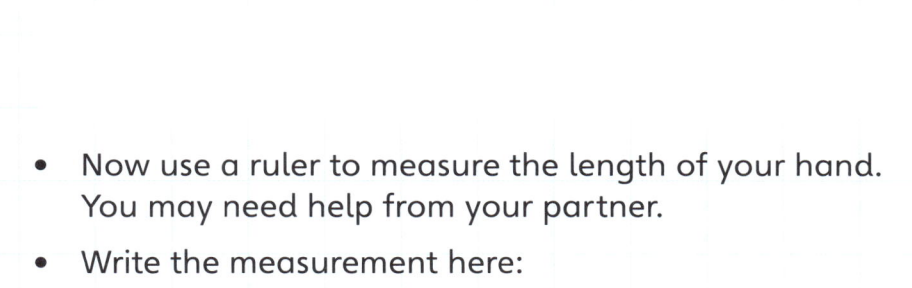

- Now use a ruler to measure the length of your hand. You may need help from your partner.
- Write the measurement here:

The length of my hand	The length of my partner's hand

- Use the length of your hand to estimate some things in your classroom.
- After each estimate, use a ruler to measure the item.
- Compare each measurement with your estimate.
- Complete this table to show your results:

Item	Estimate	Measurement	Difference

- Did you and your partner make sensible estimates?
- Explain why:

7D Measuring mass

Discover

Work in a group of four.

- Choose four different-sized items from the classroom.
- As a group estimate how much you think each item weighs. You need to agree on an estimate!
- As a group decide the best unit to use.
- Write the item, unit and your estimate in this table. Use decimal notation.

Item	Unit	Estimate	Actual	Difference

- Now weigh each item.
- Add this information to your table (again, use decimal notation).
- Compare the **mass** of each item with your estimate.
- Write the difference between your estimate and the actual **weight** in the table.

Did your estimating improve?

- Now order your masses from lightest to heaviest:

7D Measuring mass

Explore

Work in a group of four.

- Solve this problem:

> Claudia's mum teaches grade 1 in a school.
>
> She would like you to make some fruit with modelling clay for her students.
>
> She wants the fruit to weigh 150 g, 270 g, 420 g and 540 g.

- Make two fruits of each weight.
- Then make labels for each fruit to show their mass.
- When did you estimate in this activity?

- How did you get the correct mass?

The scales below weigh items up to 1 kg.

- Label them with the weight of each fruit:

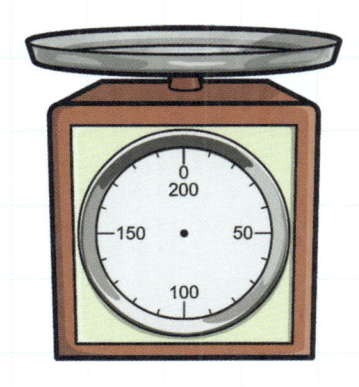

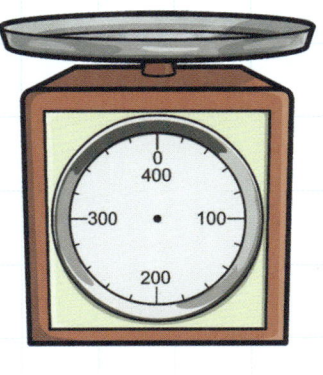

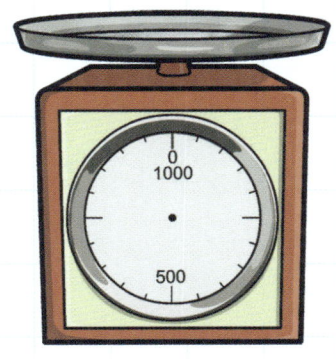

- What is the total weight of all the fruit?
- Write this in three different ways:

7E Measuring capacity

Discover

Work in a group of four.

- Choose four different **containers** from the classroom.
- As a group estimate the **capacity** of each container. You need to agree!
- As a group decide the best unit of measure to use.
- Write the item, unit and your estimate in this table.

Item	Unit	Estimate	Actual	Difference

- Now fill each container with water to find its actual capacity.
- Add this information to your table. Use decimal notation.
- Compare the capacity of each container with your estimate.
- Write the difference between the capacities and your estimates in the table.

Did your estimating improve with practice?

- Now order the capacities. Start with the smallest:

7E Measuring capacity

Explore

Work with a partner.

- Pick three digit cards.
- Use the digit cards to make a 3-digit number.

This number is the capacity of a container in millilitres.

- Write your capacity here:

  _____

- Now use the cards to make three more capacities. Write them here:

  _____

  _____

  _____

- Estimate the capacities into a container.
- Now use a measuring jug/cylinder to measure your estimates.

Capacity	Estimate amount	Difference between your estimate and the capaity

How close were your estimates?

- Show the capacities on these measuring jugs:

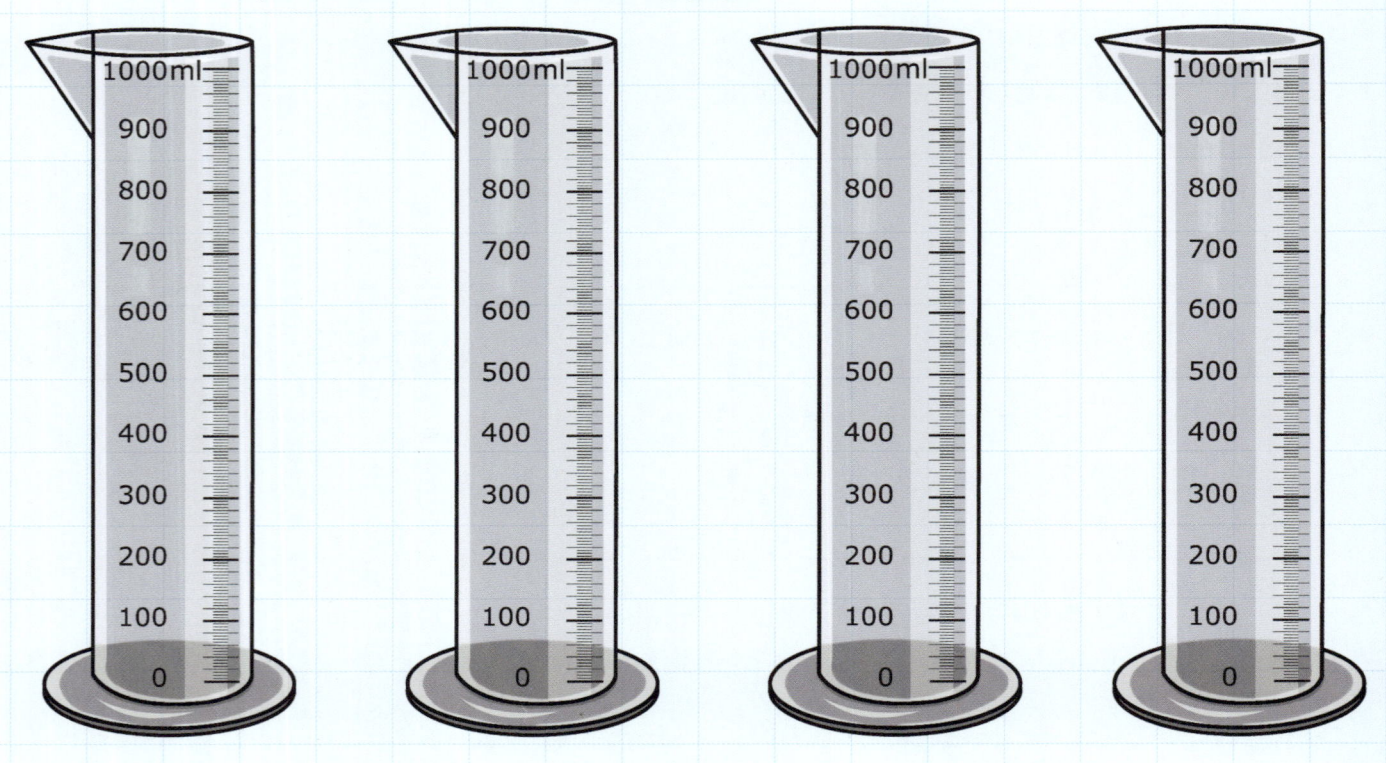

What is the total capacity of all four cylinders?

- Write this in three ways:

- Now order your capacities. Start with the smallest:

7 Length, mass and capacity

Connect

Work as a group.

Abrahim Bake wants to make a perfect chocolate cake, but he is finding it difficult.

Here is part of the instructions he is using:

> Ingredients:
> $\frac{3}{4}$ kg flour
> 2 eggs
> $\frac{1}{4}$ kg butter
> $\frac{1}{2}$ litre milk
> $\frac{1}{10}$ litre of plain yoghurt
> $\frac{1}{10}$ kg sugar
> $\frac{1}{5}$ kg chocolate powder

What is the problem with the instructions?

- Write what you think here:

    The list uses _____

    The marks on the scales are in _____

- Rewrite Abrahim's instructions for him so that he can successfully make his chocolate cake:

    _____
    _____
    _____
    _____
    _____
    _____
    _____

7 Length, mass and capacity

Review

Work with a partner.

What have you learnt about length, mass and capacity?

- Write your information on the mind map below.
- Remember to write the vocabulary you know too.
- Give examples to show what you know.

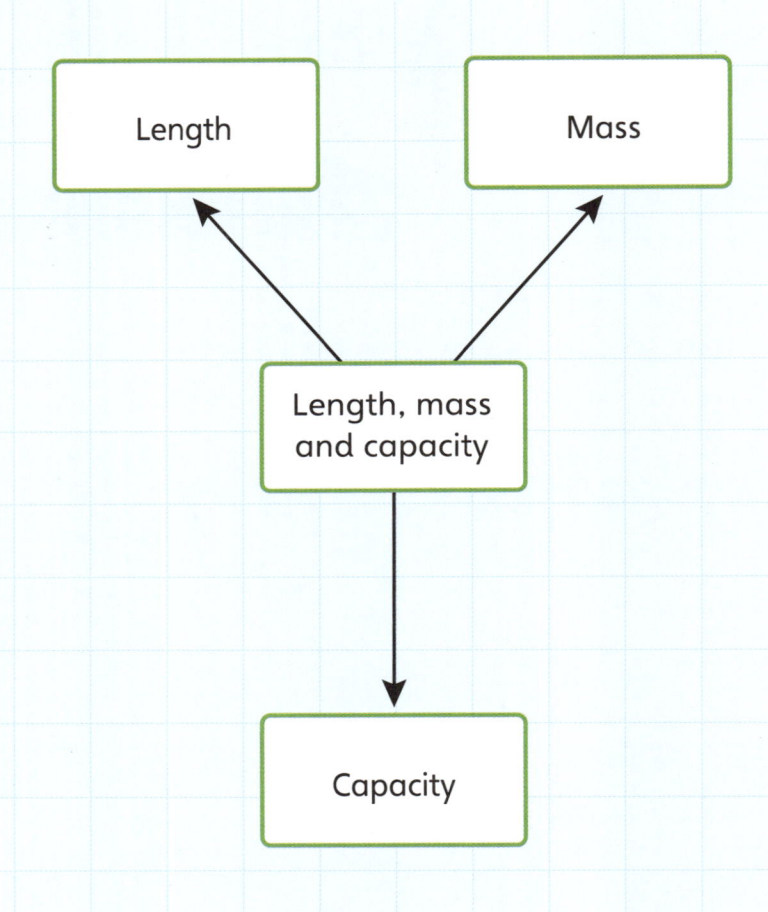

```
    Length              Mass
       ↖              ↗
      Length, mass
       and capacity
            ↓
         Capacity
```

8 Time

Engage

- Work in groups of four.
- Give the meaning for each word in the table. The rest of your group need to guess which word you are describing.
- Tick ✓ the words when they guess correctly.

millennium	month	minute
century	week	second
decade	day	analogue clock
year	hour	digital clock
yesterday	today	tomorrow
12-hour time	24-hour time	soon

- Play the game twice.
- Now write your descriptions under the words.

8A Converting between units of time

Discover

- Write the equivalent number of **seconds** for each of these **minutes**.
- Double, halve or add to work out the new facts.
- Two are shown for you. Start by halving these.

4 minutes = 240 seconds

10 minutes = 600 seconds

2 minutes = _____ seconds

5 minutes = _____ seconds

8 minutes = _____ seconds

How many seconds?

20 minutes = _____ seconds

12 minutes = _____ seconds

15 minutes = _____ seconds

27 minutes = _____ seconds

- Now do the same for these months and years.

4 years = 48 months

10 years = 120 months

2 years = _____ months

5 years = _____ months

8 years = _____ months

How many months?

20 years = _____ months

12 years = _____ months

15 years = _____ months

27 years = _____ months

8A Converting between units of time

Explore

- Work out how many hours and minutes there are in the minutes listed in this table. The first one has been done for you as an example.

	Hours	Minutes
65 minutes	1	5
90 minutes		
80 minutes		
100 minutes		
115 minutes		
125 minutes		
185 minutes		
210 minutes		
240 minutes		
265 minutes		
300 minutes		

- Now add together the minutes in this table.
- Then change them to hours and minutes in the last column: An example is shown for you.

	Minutes	Hours and minutes
50 minutes + 25 minutes	75 minutes	1 hour 15 minutes
45 minutes + 40 minutes		
50 minutes + 80 minutes		
85 minutes + 75 minutes		
32 minutes + 90 minutes		
9 minutes + 125 minutes		

8B Using the 24-hour clock

Discover

- Draw lines to match the **analogue**, **digital** and **clock times**.
  The first one has been done for you as an example.

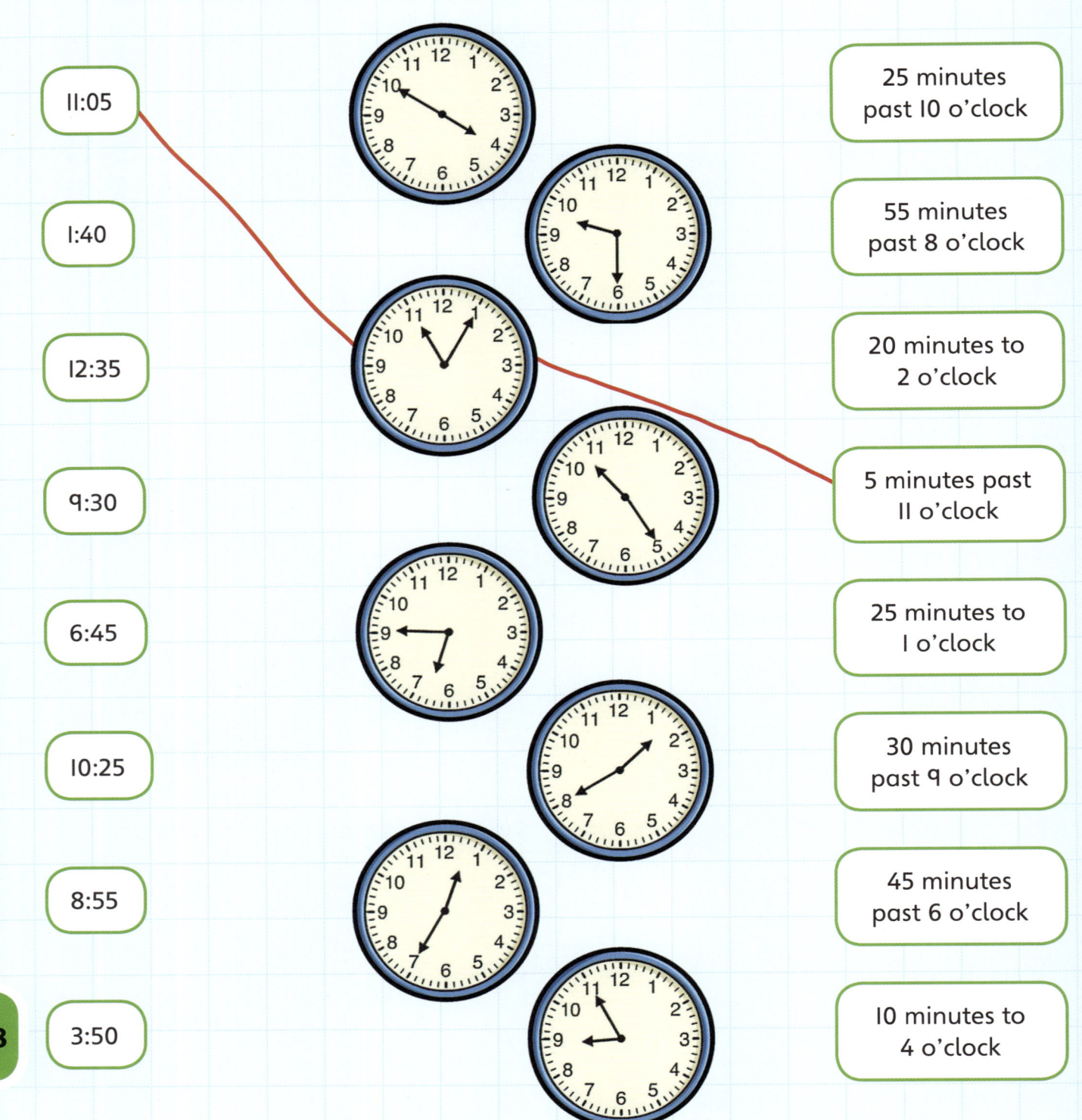

11:05

1:40

12:35

9:30

6:45

10:25

8:55

3:50

25 minutes past 10 o'clock

55 minutes past 8 o'clock

20 minutes to 2 o'clock

5 minutes past 11 o'clock

25 minutes to 1 o'clock

30 minutes past 9 o'clock

45 minutes past 6 o'clock

10 minutes to 4 o'clock

8B Using the 24-hour clock

Explore

- Solve these problems.
- Use the number lines to help you.

1. Yukesh went for a run.

   He left his house at 14:20.
   He ran for 2 hours 15 minutes.
   When did he get home?

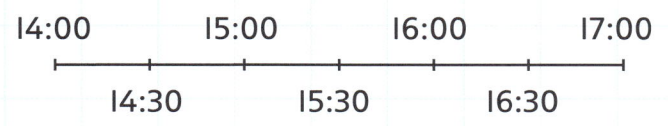

   14:00　　　15:00　　　16:00　　　17:00
   　　　14:30　　　15:30　　　16:30

   Yukesh got home at _____.

2. Nafisat spent 1 hour and 30 minutes tidying her house.

   She finished at 12:45.
   What time did she start?

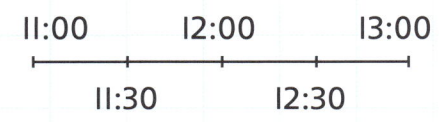

   11:00　　　12:00　　　13:00
   　　　11:30　　　12:30

   Nafisat started at _____.

3. The film started at 14:10 and finished at 16:25.

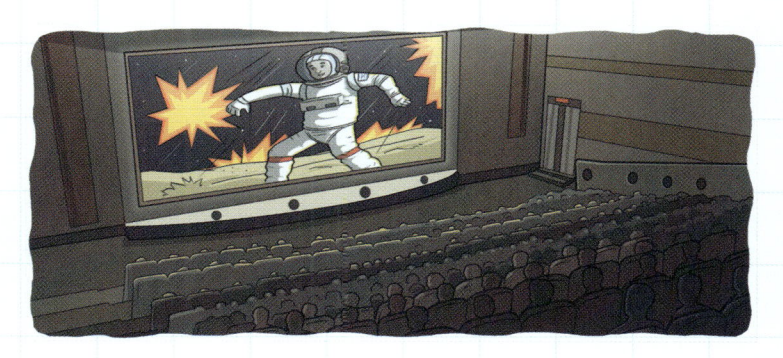

   How long was the film?

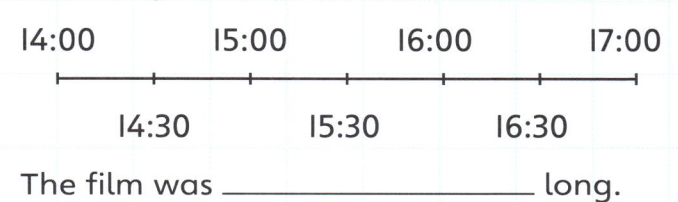

   14:00　　　15:00　　　16:00　　　17:00
   　　　14:30　　　15:30　　　16:30

   The film was _____ long.

4. Sam and Sergie watched their favourite programme on the television.

   The programme started at 15:05 and finished at 17:40.
   How long was the programme?

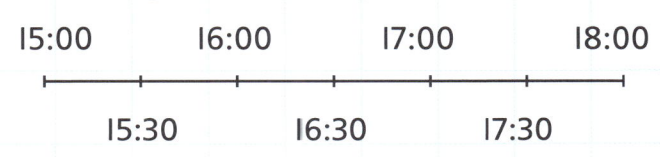

   15:00　　　16:00　　　17:00　　　18:00
   　　　15:30　　　16:30　　　17:30

   The programme was _____ long.

5. Naomi and Fatima began their homework at 15:10.

Naomi finished her homework after an hour.

Fatima finished her homework at 16:55.

How much longer did it take Fatima to do her homework than Naomi?

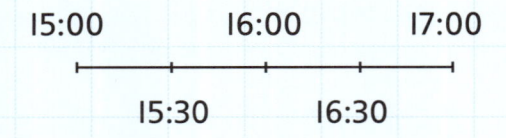

15:00   16:00   17:00
       15:30   16:30

It took Fatima _____ longer to do her homework.

- Now make up and solve a problem of your own:

8C Reading timetables

Discover

This **timetable** show the times that different buses leave the bus garage and arrive at the school:

	Bus 1	Bus 2	Bus 3	Bus 4	Bus 5	Bus 6
Bus garage	08:00	08:45	09:05	09:15	10:21	10:47
Shopping centre	08:20	09:00	09:35	09:30	10:47	10:52
Swimming pool	08:45	09:30		10:00		11:38
Park	09:05	10:00	10:20	10:30	12:06	12:19
Airport	09:55		10:55	11:00	12:42	12:53
Football stadium	10:15		11:35	11:15		13:11
School	10:45	11:45	12:20	12:00	13:36	13:52

- Work out the total time it takes for each bus to travel from the bus garage to the school. Use number lines to help you.

Bus 1

Bus 2

Bus 3

Bus 4

Bus 5

Bus 6

8C Reading timetables

Explore

This table shows the departure times from Mumbai International Airport and the arrival times at Bangkok Airport.

How long are the flights?

- Draw number lines on paper to help you.

Flight	Depart Mumbai International	Arrive Bangkok Airport	Journey length
1	0100	0850	0100 + 7 hours = 0800. Length = 7 hours 50 minutes
2	0200	0745	
3	0405	1455	
4	0655	1250	
5	0830	1725	
6	0910	1610	
7	1040	1720	
8	1155	1905	

Imran's father needs to meet some people at 16:00 in Bangkok.

He doesn't like flying so he wants the shortest flight. Which flight is best for him?

The best flight for him is Flight _____ because

_____

Why do you think some of these flights are much longer than others?

I think some flights are longer because _____

_____

8D Calculating time intervals

Discover

- Solve these problems.
- Draw number lines to help you.

1. Levi went apple picking.

   He began at 14:20.

   He picked apples for 2 hours and 25 minutes.

   What time did he finish?

   Levi finished at _____

2. Kieron spent 2 hours and 5 minutes at the gym.

   He left the gym at 18:20.

   When did he arrive at the gym?

   Kieron got there at: _____

3. The twins went to the park.

   They arrived at 15:50 and left at 17:35.

   How long were they at the park for?

   The twins were at the park for:

   _____

4. Carrie and Mona left home to go to school at 07:15.

   They spent the day working hard.

   They got home at 16:25.

   How long were they away from home?

   Carrie and Mona were away from

   home for: _____

5. Adam and Chris worked in the garden.

   They started at 14:25.

   Adam finished at 16:10.

   Chris continued working for another hour and 10 minutes.

   For how long was Chris working in the garden?

   Chris was working in the garden for

   _____

Time

8D Calculating time intervals

Explore

- Work with a partner.
- Pick three digit cards.
- Make a time using your cards. Use one card to represent the hour number and two cards to represent the minutes.
- Draw the time on the first analogue clock (in the row below the example) and write the 12-hour digital time underneath.
- Is it a morning or an afternoon time? Write **a.m.** or **p.m.** beside the time.

For example: Cards picked: 5 4 9

Digital time: 9:45

It is morning, so the time is 9:45 a.m.

- Repeat the activity three times, choosing a new set of three cards each time.

Cards picked: Cards picked: Cards picked: Cards picked:

Digital time: Digital time: Digital time: Digital time:

8E Using calendars

Discover

January	February	March	April
S M T W T F S	S M T W T F S	S M T W T F S	S M T W T F S
1 2 3 4	1	1	1 2 3 4 5
5 6 7 8 9 10 11	2 3 4 5 6 7 8	2 3 4 5 6 7 8	6 7 8 9 10 11 12
12 13 14 15 16 17 18	9 10 11 12 13 14 15	9 10 11 12 13 14 15	13 14 15 16 17 18 19
19 20 21 22 23 24 25	16 17 18 19 20 21 22	16 17 18 19 20 21 22	20 21 22 23 24 25 26
26 27 28 29 30 31	23 24 25 26 27 28	23 24 25 26 27 28 29 / 30 31	27 28 29 30

2014

- Mark your birthday on the **calendar**.
- Now mark on another important date.

  For example:
  - the birthday of a family member or friend
  - a special celebration.

What is the length of time between the two dates?

- Do this again for two other dates.

What is the length of time between these two dates?

8E Using calendars

Explore

- Follow this date trail.
  It leads to Adnaan's birthday.

F means go forward.

B means go back.

- Find the quickest way to move around the calendar.
  For example: instead of counting 21 days, move along 3 weeks:
  - Start on 2 February
  - F 35 days
  - B 72 hours
  - F 48 hours
  - F 15 days
  - B 96 hours
  - F 22 days
  - B 24 hours
  - F 30 days
  - B 8 days

- Use this space for your calculations.

You have now reached Adnaan's birthday!
When is his birthday?

8 Time

Connect

- Work as a group.
- Take a survey of the favourite television programmes of the class.
- Pick the four most popular.
- Make up a TV guide for these programmes.

You need to agree the start time and the finish time.

Make sure that each programme follows on from the previous one.

Each programme needs to last a different amount of time from the others.

- Work out how long each programme lasts.
- Write all this information in this table.
  Two examples are shown.

TV programme	Start time	Finish time	Length of programme
The Pirate's Treasure Hunt	16:00	16:25	25 minutes
On Safari	16:25	17:35	1 hour 10 minutes

8 Time

Review

- With your partner, think of all you have been learning about time.
- Under each heading give examples to show what you know.

You can draw pictures and use words.
For example: in the space for analogue clock times you can draw a clock face and write 6 o' clock.

You can look in your book to remind you.

How confident are you at solving problems involving 12-hour and 24-hour clock times?

I am:

Analogue clock times	12-hour clock times

Time intervals	Units of time

Equivalent units of times	Timetables

138

9 Perimeter and Area

Engage

- Work in a group.
- Each of you draws one or two squares on cm-squared paper. Each square needs to have sides of 10 cm.
- Your group needs five squares altogether.
- Cut out your squares.
- As a group use your five squares to make eight different shapes. Use all five squares for each shape.
- Draw your shapes here:

- Together talk about these questions:
  - Do you think the **areas** are the same?
  - Why do you think that?
  - Do you think the **perimeters** are the same?
  - Why do you think that?
- Now work out the area of each shape.
- Label your drawings with the areas.
- Now work out the perimeters of each shape.
- Label your drawings with the perimeters.

Were you and the rest of your group correct?

9A Understanding perimeter

Discover

- Work with a partner.
- Draw three rectangles in this space.
  The lengths of the rectangles need to be in whole centimetres.

- Now give your rectangles to your partner.
- Your partner needs to estimate their perimeters.
  They write the perimeters inside your shapes.
- Your partner can then work out the exact perimeters, using a ruler and one of the formulae. They then write the exact perimeters inside your shapes.
- Swap books again.
- Now check: How close were your partner's estimates?

1st rectangle: _____

2nd rectangle: _____

3rd rectangle: _____

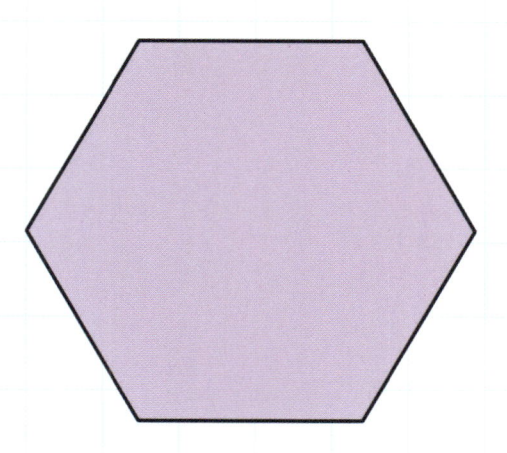

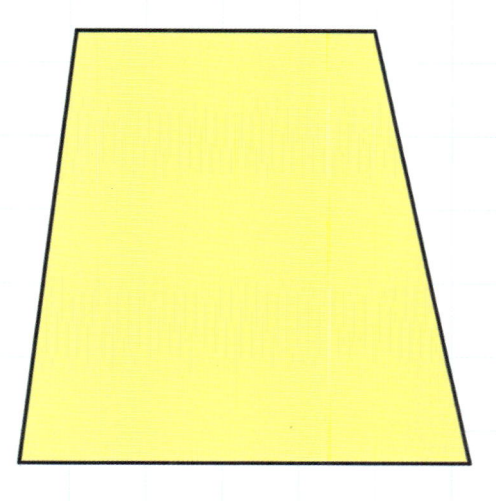

9A Understanding perimeter

Explore

- Measure the sides of each shape.
- Label the sides with the measurements.
- Work out the perimeter of each shape.
- Use a formula when you can.

Perimeter:

Perimeter:

Perimeter:

Perimeter:

Perimeter:

9B Understanding area

Discover

Harry thinks that when the area of a shape gets bigger so does the perimeter.

What do you think?

Is what Harry thinks:
- always true?
- sometimes true?
- never true?
- Use this space to explain your thinking:

- Draw squares and rectangles on cm-squared paper as part of your explanation. You can stick them here:

9B Understanding area

Explore

- Find the areas of these shapes.
- Count the squares in the quickest way you can.
- Use the formula discussed in class to check.

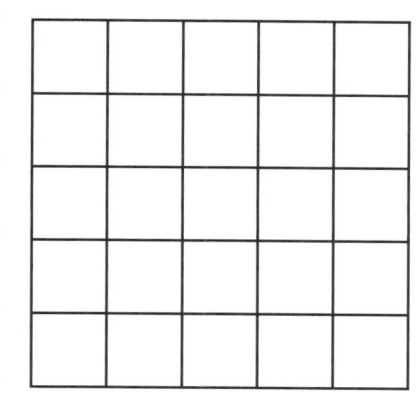

The area of this shape is:
_____
This is how I worked it out:

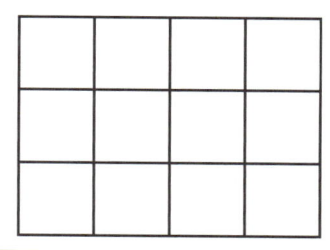

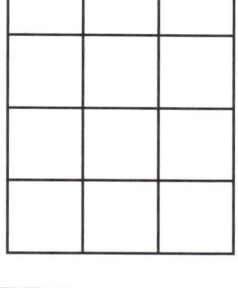

The area of this shape is:
_____
This is how I worked it out:

The area of this shape is:
_____
This is how I worked it out:

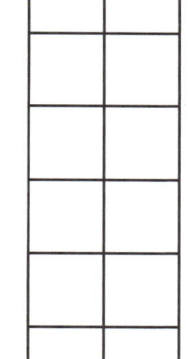

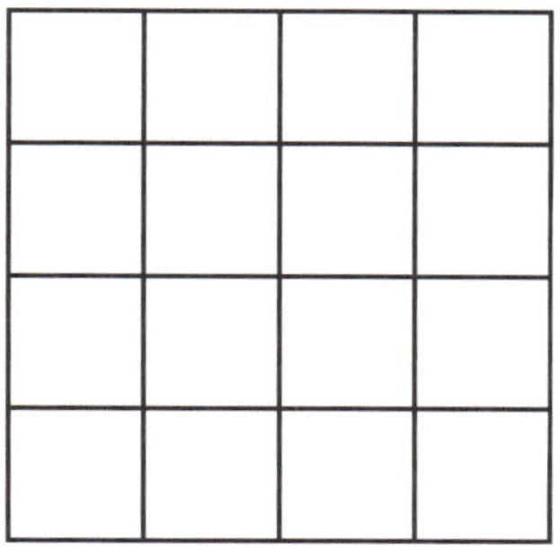

The area of this shape is:
_____
This is how I worked it out:

The area of this shape is:
_____
This is how I worked it out:

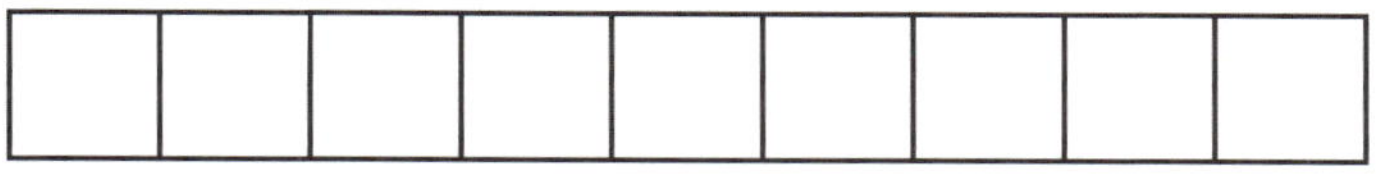

The area of this shape is:
_____
This is how I worked it out:

The area of this shape is:
_____
This is how I worked it out:

9C Calculating areas and perimeters of rectangles

Discover

Sophie wants to build a rectangular patio in her garden.
She wants the area of her patio to be 24 m².

- What are the possible sizes for Sophie's patio? Write these here:

- Draw some designs using these sizes. Use a scale of 1 cm = 1 m. (Use a separate piece of paper if you need more space.) Measure accurately using your ruler.
- Label the measurements.
- Check your drawings to make sure the areas are correct.
- Write the perimeters of each shape.

For example:

2 m

12 m   NOT TO SCALE

Perimeter = 28 m

Which one of your designs do you think is the best for Sophie to use? Why?

I think the best design for Sophie to use is _____

because _____

9C Calculating areas and perimeters of rectangles

Explore

- First estimate the perimeter and area of these shapes.
  Write your estimates in the spaces provided beside each shape.
- Then use a ruler to measure the perimeter.
- Count the squares inside the shapes to find the area. Each square represents $1 cm^2$.
- How accurate were your estimates?

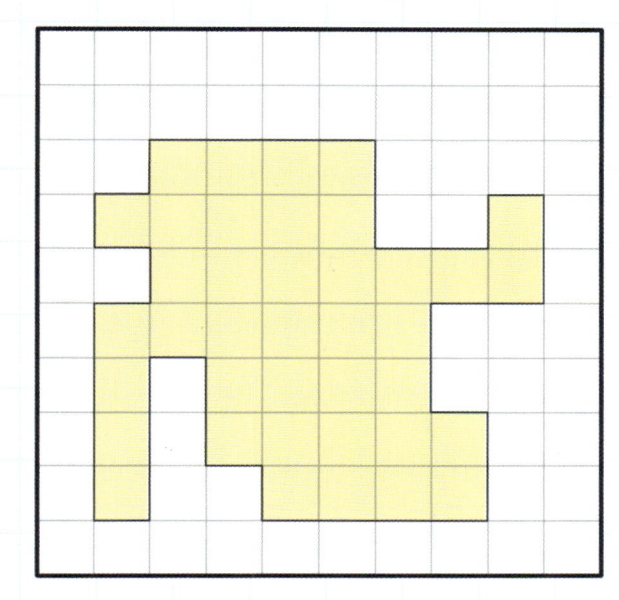

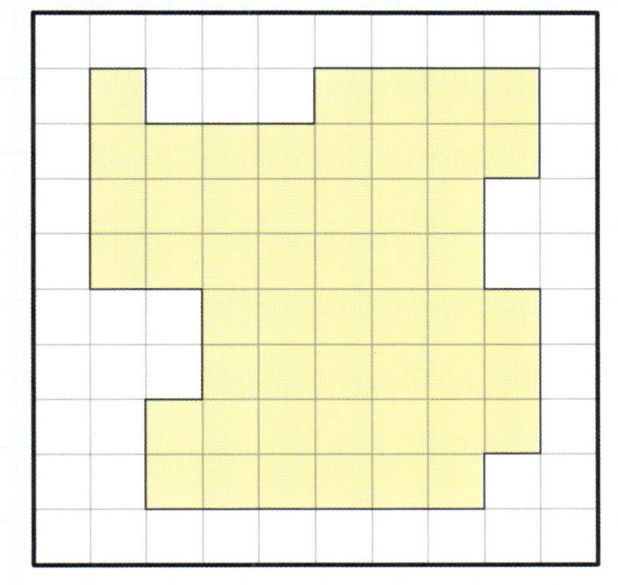

- My estimate of the perimeter is _____
- The actual perimeter: _____
- Difference: _____
- My estimate of the area is _____
- The actual area: _____
- Difference: _____

- My estimate of the perimeter is _____
- The actual perimeter: _____
- Difference: _____
- My estimate of the area is _____
- The actual area: _____
- Difference: _____

9 Perimeter and area

Connect

Fatima has a large area of land. She wants to build a stable for her horse on part of it.

She wants the stable to be a rectangular shape with a perimeter of 50 m.

- Work in a group.
- What are some possible areas for Fatima's stable? Write these here:

- On paper, sketch some designs using your areas. (Your measurements do not need to be to scale.)
- Remember to label the area on each design.
- Check your drawings to make sure the perimeters are correct.
- Write the areas of each shape.

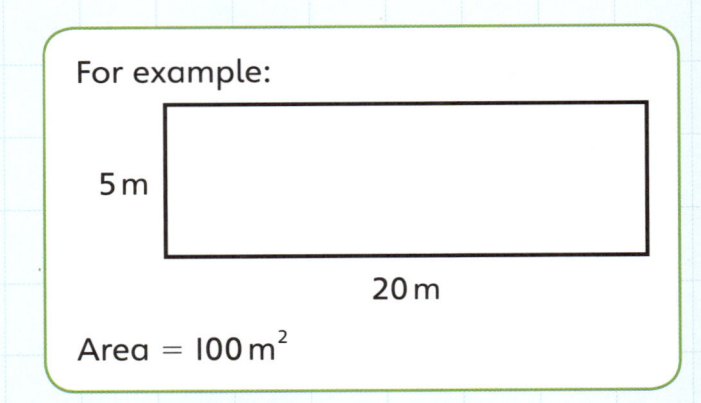

For example:

5 m

20 m

Area = 100 m^2

Which one of your designs is the best for Fatima to use? Explain why:

I think the best design for Fatima to use is _____

because _____

_____

- Draw your design here:

9 Perimeter and area

Review

- Draw two squares and two rectangles in this space.
- Measure their lengths and widths in whole centimetres. Make sure you are accurate when you use your ruler.
- Label your shapes with these measurements.

- Now work out the perimeters of your shapes.
  Use one of the formulae you learned about during this Unit.
- Write the perimeter beside each shape.
- Now work out the areas of your shapes.
  Use the formula l × w (length × width).
- Write the area inside each shape.

10 Handling Data

Engage

Look at these ways of representing information. What are they called?
- Write their names.

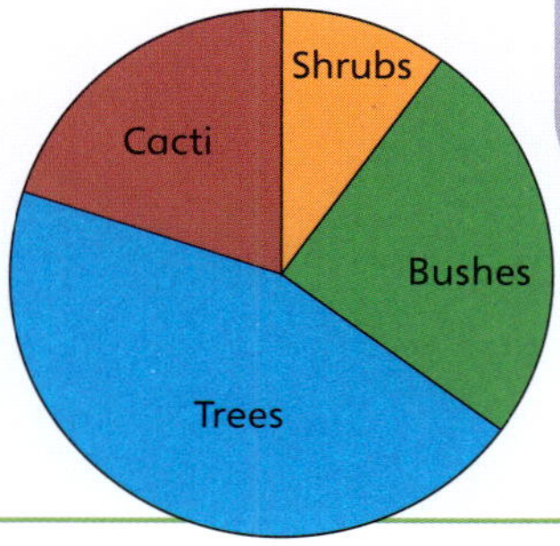

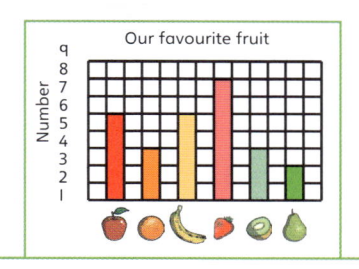

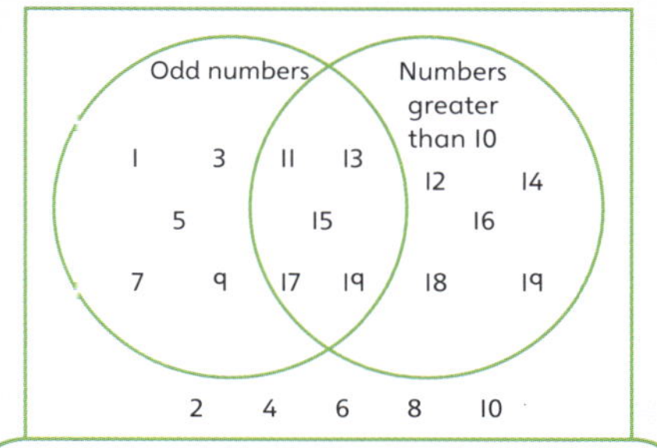

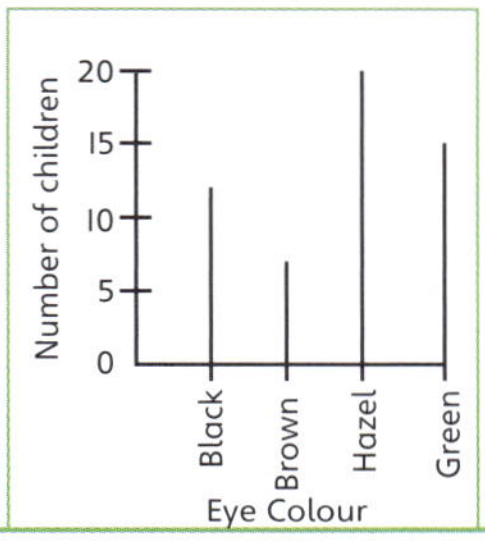

What do they tell us?
- Discuss this with your group.

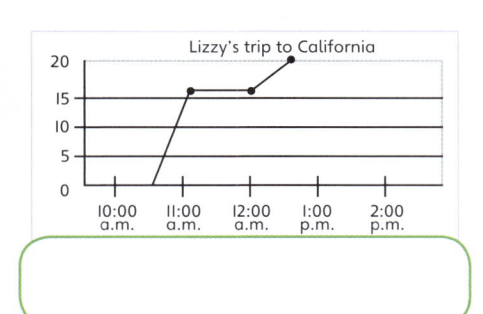

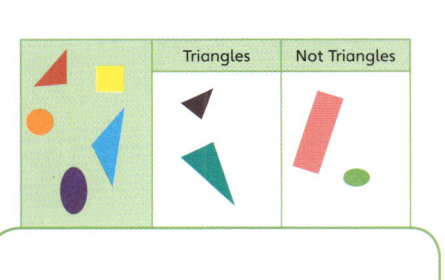

Fruit	Tally number
Papaya	|||| |||
Blueberry	||||
Apple	||||	
Strawberry	|||

Colour of bike	Tally	Frequency
Red	|||| |||| ||	12
Yellow	||||	5
Green	|||| ||||	9
Blue	||||	4
Orange	|||| |||	8

149

10A Frequency tables and pictograms

Discover

Think of something you want to find out about your class.
Write it as a question below.

How can you find out the information to answer your question?

- Write down the question you might ask.
- Draw a frequency table below which records the results of your survey.

Write down 3 facts you know from your table.

10A Frequency tables and pictograms

Explore

Snack	Frequency
Mango	8
Apple	6
Carrot	3
Energy bar	9
Potato chips	5

Mango	👤 👤 👤 👤
Apple	👤 👤 👤
Carrot	👤 ⌇
Energy bar	👤 👤 👤 👤 ⌇
Potato chips	👤 👤 ⌇

Use the information in the frequency table from the **Discover** lesson to draw a pictogram.

Draw a pictogram using data from another group in the space below.

Write 3 sentences about the data in this pictogram.

10B Bar line graphs and line graphs

Discover

- Work in a small group.
- Think of something you would like to find out about the class – for example: their favourite animal, sport or food.

You could choose one of these ideas but, better still, think of one of your own.

How will you collect your **data**?

- Collect your data.
- Represent it in a **bar line graph**.

Use a **vertical axis** that goes up in intervals of 2.

- Draw your graph here:

- Write some questions on another piece of paper to ask the class about your graph.

10B Bar line graphs and line graphs

Explore

- Look carefully at this **line graph**.
  It shows 12 hours in the life of truck driver Pete.

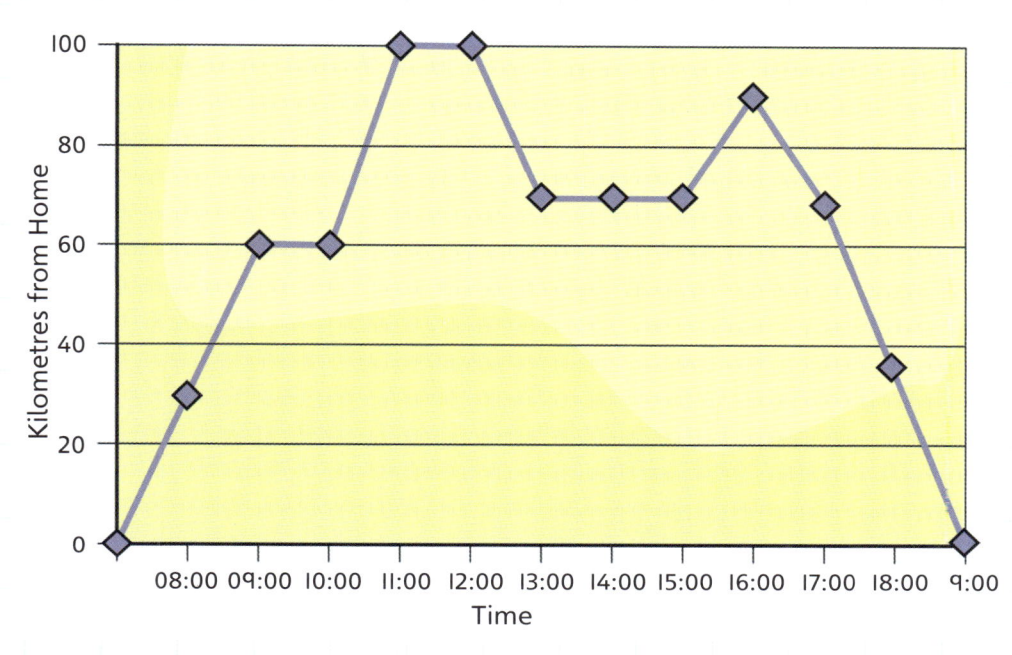

1. How far from home is Pete at 09:00?

2. What do you think he is doing from 09:00 to 10:00?

3. How many kilometres does Pete travel from 10:00 to 11:00?

4. For how many hours does he not travel anywhere?

5. How far from home is he at 13:00?

6. How far from home is he at 16:00?

7. Estimate how far from home he is at 17:30.

8. How many kilometres did Pete travel in total?

Handling Data

153

10C Line graphs

Discover

This data shows temperatures over a 12-hour period:

12:00	13:00	14:00	15:00	16:00	17:00	18:00	19:00	20:00	21:00	22:00	23:00
24°C	25°C	25°C	24°C	24°C	23°C	23°C	22°C	21°C	20°C	17°C	17°C

- Work with a partner.
- On squared paper draw a line graph to show this data.

What is the **mode** of the data?

- Estimate what the temperature was at:
  - 19:30
  - 13:15
  - 17:45
  - 22:30
  - 12:15
  - 19:45
- Now make up some questions to ask the class:

10C Line graphs

Explore

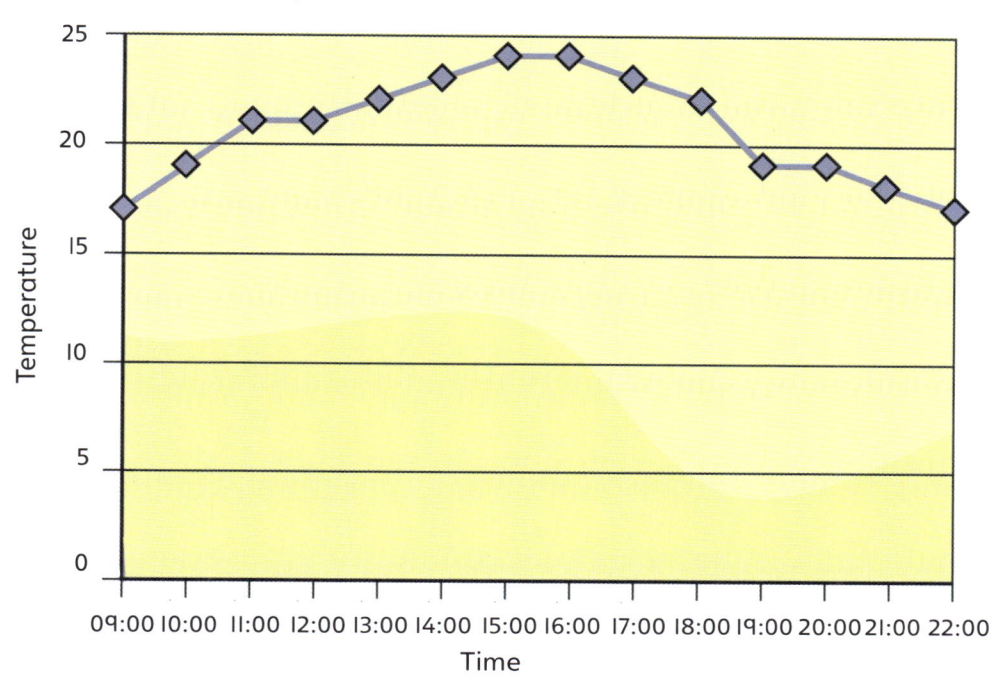

Temperatures over 12 hours in Nairobi

This line graph shows the temperatures over 12 hours in Nairobi, Kenya.

- Look carefully at the graph.

    1. What is the temperature at 14:00?

    2. What is the temperature at 17:30?

    3. George wants to go for a run during the two coolest hours of the day. When is the best time?

4. Daisy likes to be outside during the hottest part of the day. Which four hours are best?

5. What is the mode temperature?

- Make up some questions from this graph to ask the class:

    1. _____
    2. _____
    3. _____
    4. _____

Handling Data

155

10D Probability

Discover

Complete this table by writing events in the spaces.
Add two phrases of your own.

Likelihood	Event
Impossible	
Very unlikely	
Unlikely	
Even chance	
Likely	
Very likely	
Will definitely happen	

Pick an answer you are sure about. How do you know?

Pick an answer you are uncertain about. Explain why.

10D Probability

Explore

Complete this table after your experiment.

Side of coin	Frequency

Write down the likelihood of a coin falling on the side you guess.

Complete this table after your experiment.

Number	Frequency
1	
2	
3	
4	
5	
6	

Write two sentences about the probability of getting a given number when you roll a dice.

10 Handling data

Connect

- Work as a group.

This frequency table shows the medals won in the 2012 Olympics by the top ten countries:

2012 Summer Olympics medal table					
Rank	Country	Gold	Silver	Bronze	Total
1	USA	46	29	29	
2	China	38	27	23	
3	Great Britain	29	17	19	
4	Russia	24	26	32	
5	South Korea	13	8	7	
6	Germany	11	19	14	
7	France	11	11	12	
8	Italy	8	9	11	
9	Hungary	8	4	6	
10	Australia	7	16	12	

- Together, work out the total medals won by each country.
- On a large piece of paper put information from the table on a bar line graph.

Decide which information to use.

For example: the total medals each country has won, or the total gold, bronze or silver medals won.

Remember to consider the scale on the vertical axis.

What is the best interval to use?

Give your graph a title and label each axis.

- Now make up some statements from your bar line graph.

- Write your statements here:

10 Handling data

Review

- Look carefully at these ways of representing data.
- Beside each one, write its name.
- Then write what information it gives.
- Next write something that it tells you.

Colour of Car	Tally	Frequency
Orange	|||| ||||	10
Yellow	||||		6
Red	|||| |||	8
Blue	||||	4
Green	|||| |||	8
Purple	|||| ||||		11

This is a _____

It shows _____

One piece of information that it

gives is _____

_____

_____

Apple
Orange
Mango
Banana
Grapes
Pineapple

🍎 = 7

This is a _____

It shows _____

One piece of information that it

gives is _____

_____

_____

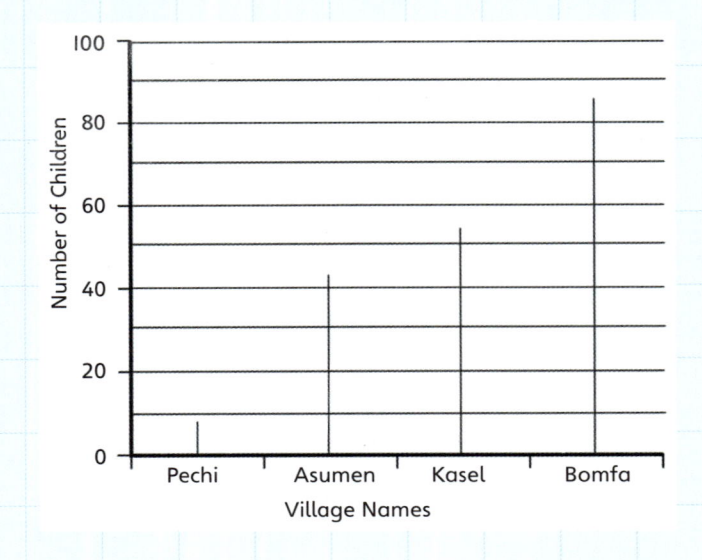

This is a _____

It shows _____

One piece of information that it

gives is _____

_____

_____

This is a _____

It shows _____

One piece of information that it

gives is _____

_____

_____

160

Glossary

12-hour clock

_____
_____

4.30 p.m.

24-hour clock

_____
_____

16:30 is 4.30 in the afternoon.

acute angle

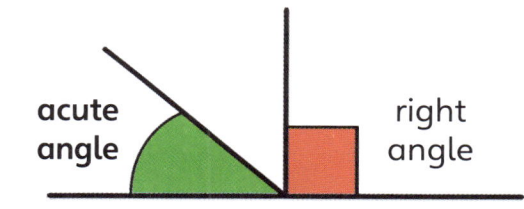

_____
_____
_____

acute angle   right angle

approximately equal to (≈)

_____
_____

ascending order

_____
_____

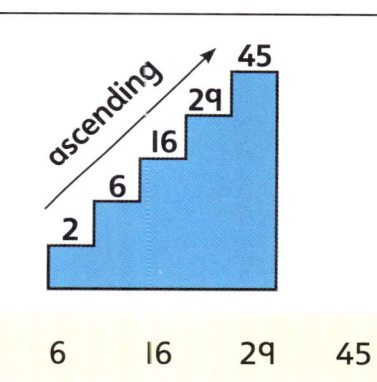

2   6   16   29   45

These numbers are in **ascending order**.

axis of symmetry

_____
_____

axis of symmetry

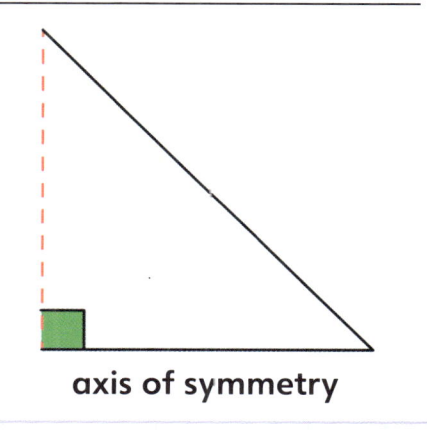

bar line chart

_____

_____

_____

Visitors to the museum

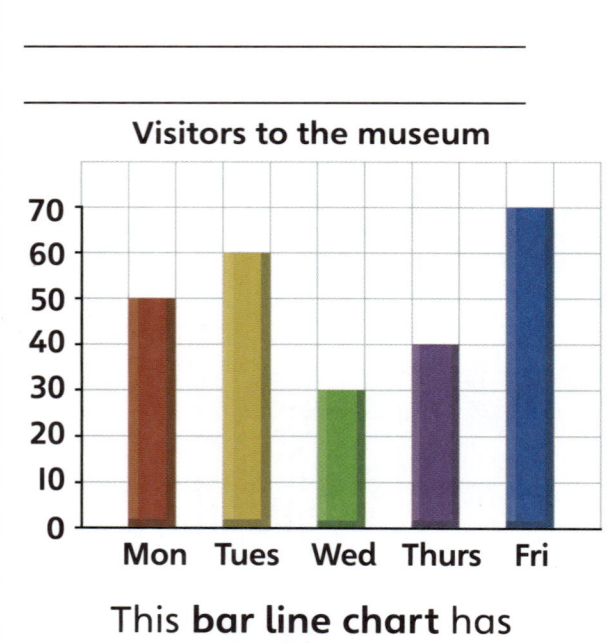

This **bar line chart** has vertical bar lines.

cancel

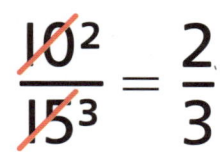

_____

_____

_____

_____

$$\frac{\cancel{10}^{2}}{\cancel{15}^{3}} = \frac{2}{3}$$

The numerator and denominator have both been divided by 5.

$\frac{10}{15}$ has been **cancelled** to $\frac{2}{3}$.

bisect

_____

_____

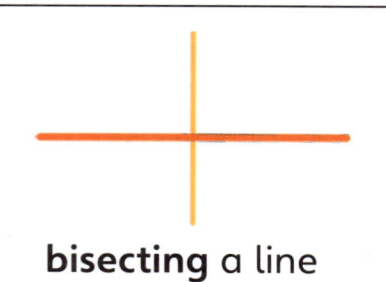

bisecting a line

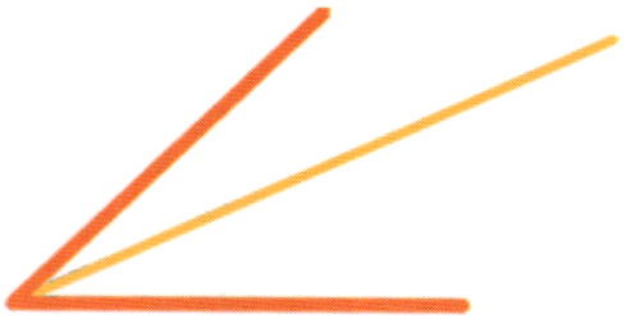

bisecting an angle

certain

_____

_____

_____

_____

chance
(good/poor/no chance)

_____

_____

_____

congruent

_____
_____
_____

constant

_____
_____
_____

convert

_____
_____
_____

120 cm **converts** to 1.2 m.
0.55 kg **converts** to 550 g.

currency

_____
_____
_____

database

_____
_____
_____

A **database** stores information.

denominator

_____
_____
_____

numerator
$$\frac{2}{3}$$
denominator

The **denominator** tells us that this fraction is **thirds**.

descending order

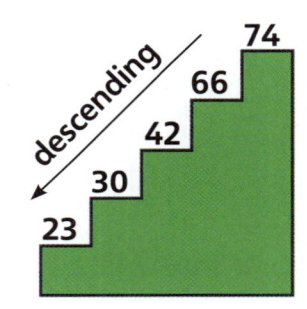

_____
_____
_____

74 66 42 30 23

These numbers are in **descending order**.

discount

_____
_____
_____
_____

Full price $ 250
Discount price $ 199
You save $ 51

divisibility rules

_____
_____
_____

Some **divisibility rules**:
- All even numbers are divisible by 2.
- All numbers that end in 5 or 0 are divisible by 5.

equivalent

_____
_____
_____
_____

Equivalent fractions:

$\frac{4}{6} = \frac{2}{3}$ $\frac{3}{4} = \frac{75}{100}$

Simplifying makes an **equivalent** fraction: $\frac{\cancel{10}^2}{\cancel{15}^3} = \frac{2}{3}$

fair

_____
_____
_____

formula (plural: formulae)
_____
_____
_____

> **Formula** for finding the area of a rectangle:
> $A = l \times w$
>
> **Formula** for finding the circumference of a circle:
> $C = \pi \times d$

gallon
_____
_____
_____
_____

> 1 **gallon** = 8 pints
> 1 **gallon** = 4.55 litres

greater than or equal to (≥)
_____
_____
_____

hundredth
_____
_____
_____

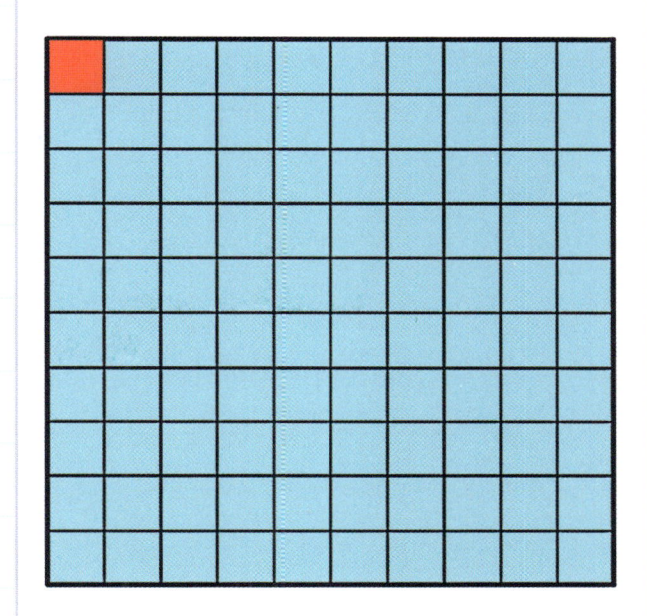

One **hundredth** of the shape is coloured.

What is one hundredth of 300?

Answer: 3

identify
_____
_____
_____
_____

impossible

_____

_____

improper fraction

_____

_____

_____

$\frac{7}{3}$ This an **improper fraction**.

less than or equal to (≤)

_____

_____

likelihood

_____

_____

likely

_____

_____

line graph

_____

_____

_____

_____

_____

maximum value

_____
_____
_____

19°C 23°C 21°C 18°C 22°C

The **maximum value** is 23°C.

This was the highest temperature recorded.

minimum value

_____
_____
_____

−1°C −3°C 0°C −2°C 2°C

The **minimum value** is 23°C.

This was the lowest temperature recorded.

mode

_____
_____
_____

Here are five numbers:

3   3   4   6   9

The **mode** is 3 because it occurs most often.

mixed number

_____
_____
_____

$2\frac{1}{2}$ is a **mixed number**.

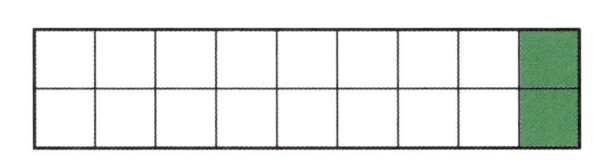

ninth

_____
_____

One **ninth** of the shape is coloured.

$\frac{1}{9}$ of 18 is 2.

numerator

_____
_____
_____

$\frac{2}{3}$ numerator / denominator — The **numerator** tells us that there are **two** thirds.

obtuse angle

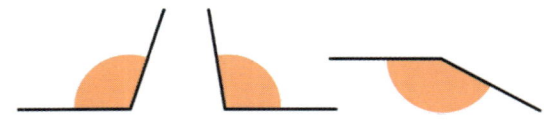

_____
_____

Obtuse angles are more than a right angle.

outcome

_____
_____
_____

octahedron

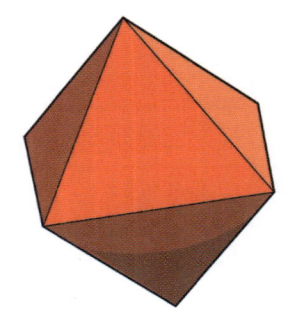

_____
_____

a regular **octahdron**

parallel

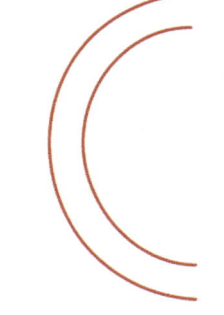

_____
_____

percentage (per cent, %)

_____

_____

Percentages can be written as fractions.

$50\% = \frac{50}{100} = \frac{1}{2}$

$25\% = \frac{25}{100} = \frac{1}{4}$

$10\% = \frac{10}{100} = \frac{1}{10}$

perpendicular

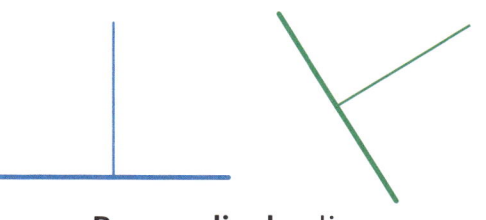

_____

_____

Perpendicular lines

probability

_____

_____

_____

_____

Toss a coin.
The **probability** of getting heads is 1 in 2.
The **probability** is $\frac{1}{2}$.

Roll a dice.
The **probability** of getting a six is 1 in 6.
The **probability** is $\frac{1}{6}$.

proper fraction

_____

_____

$\frac{7}{8}$ This a **proper fraction**.

protractor

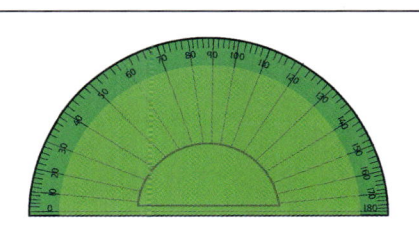

_____

_____

_____

a protractor

quadrant

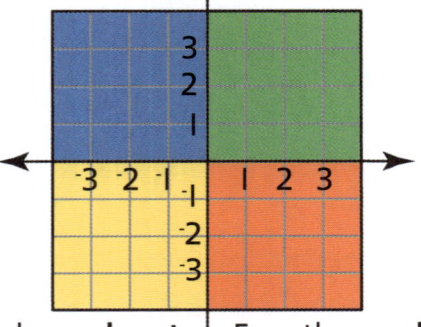

_____
_____
_____
_____

Second **quadrant** | First **quadrant**
Third **quadrant** | Fourth **quadrant**

range

_____
_____
_____
_____

Here are five numbers:

3   3   4   6   9

The smallest number is 3, the largest number is 9.

The **range** is from 3 to 9, which is 6.

ratio

_____
_____
_____

The **ratio** of cars to bicycles is 3:2.

reduced to

_____
_____

$\frac{8}{12}$ can be **reduced to** $\frac{2}{3}$.

$\frac{6}{15}$ can be **reduced to** $\frac{2}{5}$.

reflective symmetry

_____
_____

The flower has **reflective symmetry**.
The mirror line is the line of symmetry.

scalene triangle

A **scalene triangle** can be acute or obtuse. All the lengths of sides and angles are different.

square metre

1 **square metre** = 10 000 square centimetres

1 **m²** = 10 000 cm²

square millimetre

A 1 cm square has an area of 100 **mm²**

100 **mm²** = 1 cm²

square number

$1 \times 1 = 1$ $2 \times 2 = 4$ $3 \times 3 = 9$
$4 \times 4 = 16$ $5 \times 5 = 25$
1, 4, 9, 16, 25

These are all **square numbers**.

squared (1^2, 2^2, etc.)

6 **squared** is $6 \times 6 = 36$
$6^2 = 36$
2.5 **squared** is $2.5 \times 2.5 = 6.25$
$2.5^2 = 6.25$

tenths boundary

twelfth

_____
_____
_____

One **twelfth** of the shape is coloured.

$\frac{1}{12}$ of 36 is 3.

uncertain

_____
_____
_____

units boundary

_____
_____
_____

unlikely

_____
_____
_____

x-axis

_____
_____

x–axis

y-axis

_____
_____

y–axis